mental_floss: The Book

ALSO AVAILABLE FROM **MENTAL**_FLOSS

The **mental**_floss *History of the United States*

The **mental**_floss *History of the World*

mental_floss presents: *Be Amazing*

mental_floss presents: *Condensed Knowledge*

mental_floss presents: *Forbidden Knowledge*

mental_floss presents: *In the Beginning*

mental_floss: *Cocktail Party Cheat Sheets*

mental_floss: *The Genius Instruction Manual*

mental_floss: *Scatterbrained*

mental_floss: *What's the Difference?*

mental_floss: The Book

THE GREATEST LISTS IN THE HISTORY OF LISTORY

Edited by
Ethan Trex,
Will Pearson,
and Mangesh Hattikudur

HARPER

NEW YORK . LONDON . TORONTO . SYDNEY

HARPER

HarperCollins books may be purchased for educational, business, or sales promotional use. For information please write: Special Markets Department, HarperCollins Publishers, 10 East 53rd Street, New York, NY 10022.

FIRST EDITION

Cover design by Winslow Taft
Interior design by Mike Rogalski

Library of Congress Cataloging-in-Publication Data is available upon request.

ISBN 978-0-06-206930-6

11 12 13 14 15 /RRD 10 9 8 7 6 5 4 3 2 1

WITH CONTRIBUTIONS FROM:

Scott Allen
Eric Alt
Erika Beras
Doug Cantor
Miss Cellania
Alisson Clark
Chris Connolly
Stacy Conradt
Adrienne Crezo
Paul Davidson
Caroline Donnelly
Eric Elfman
Jason English
Kelly Ferguson
Kevin Fleming
Jeff Fleischer
David Goldenberg
Dan Greenberg
Meghan Holohan
David K. Israel
Mark Juddery
Allison Keene
Maggie Koerth-Baker
Rob Lammle
Ian Lendler
Neely Harris Lohmann
Mark S. Longo
Elizabeth Lunday
Kara Kovalchik
Diane Mapes
Linda Rodriguez
McRobbie
Laurel Mills
Jim Noles
David A. Norris
Mark Peters
Jason Plautz
Ransom Riggs
Greg Sabin
Erik Sass
Terri Schlichenmeyer
Streeter Seidell
Jim Smith
Matt Soniak
Brendan Spiegel
Michael A. Stusser
Christa Weil
Steve Wiegand
Graeme Wood
Sandy Wood
Rebecca Zerzan

TABLE OF CONTENTS

10 POP CULTURE LISTS TO BREAK OUT ON THE RED CARPET

10 LISTS FOR PEOPLE WHO CAN'T WRITE GOOD

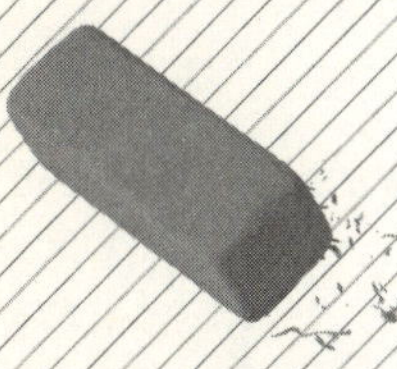

10 ANIMAL LISTS THAT DON'T BITE

10 LISTS TO READ BEFORE NAMING YOUR CHILD, COMPANY, OR ALTER-EGO

A FUNNY THING HAPPENED ON THE WAY TO THIS INTRODUCTION

I have a confession to make: *mental_floss* is not the first media start-up I've been involved in. Back when I was 9, my friends and I had a genius idea for a very different publication. It was called *Dogs of My Neighborhood!* As you've probably guessed, the magazine contained hilarious cartoons and exciting reporting on the neighborhood's most lovable mutts. The facts are a little hazy, but I think the staff spent one or two long bus rides talking about how many quarters the business would generate. Then we devoted an entire afternoon to writing, drawing, and stapling loose leaf sheets together. It was grueling work, but we believed in the product. Sadly, *Dogs of My Neighborhood!* was a commercial failure. And like many of the dogs it had once so joyously covered, it was mercifully put to sleep.

Luckily, my second stint in the media business has been more successful. As the very official "10 Outstanding Years" seal on the front of this book indicates, *mental_floss* turns ten this year. In many ways, this collection is a celebration of that milestone. But instead of boring you with the full story of how a pair of college friends started a knowledge magazine and website (you can find that stuff online!), or why you should read this book (it's great!), I thought we could instead use this space to cover how we kept this media property around for a decade. And in the spirit of this book, I'm presenting the information in list form. Here are:

4 EXCLUSIVE, RARELY REVEALED TIPS TO KEEPING YOUR START-UP AROUND

1 Don't Poke a Sleeping Big-Foot Believer

mental_floss isn't intended to offend. We try to have fun without being political or mean-spirited. So, when the magazine featured an article on Ray Wallace and his famous Big Foot hoax (in 1958, Wallace strapped 16-inch paddles to his shoes, stomped around some mud fields, then let people discover the giant footprints), we didn't expect a backlash. Instead, the hate mail poured in. Apparently, a society of Big Foot Believers had discovered our article, and in their outrage, started a campaign to get *mental_floss* to retract the statement. While we never wrote an apology, we also never spoke ill of Big Foot again. This is a good rule for any media start-up (especially those made up of scrawny wimps). But, just to be clear, Big Foot isn't the only creature on our "do not offend" list. Other no-no's include: the Loch Ness monster, the Brontosaurus, Leprechauns, and Voldemort. Yetis and unicorns were taken off the list in late 2006.

Be Careful with the *New York Times*

I don't mean plagiarism or stealing story ideas, though if you intend to run a publication, you should avoid that too. I mean, be careful where you physically place the paper! Back when *mental_floss* had an office in Birmingham, we also had a tiny, adorable office puppy named Bailey. At the time, we were trying to impress the local press, and before one particular reporter came by, we left marked-up newspapers and magazines strewn all over the conference table. Then, after we met for a while, we went to lunch. But during that break, the puppy who'd been newspaper-trained couldn't find a place to go, so he used his 3-foot vertical to leap onto the table and squat over the front page of the nation's most prestigious paper. Thankfully, we got to the mess before the reporter did. But since the pup wasn't going anywhere, and the guests weren't going to stop dropping by, we instituted a newspaper rule—namely, don't leave them in hopping distance!

Don't Try to Outsmart Google Employees

When Will and I were invited to visit Google headquarters in California, we decided to have some fun with the event. Before the talk, we handed out the World's Geekiest Crossword puzzle, where all the answers came from constructed languages: Klingon, Elvish, Esperanto, etc. Since we didn't really expect anyone to know these languages off the top of his head, we offered elaborate prizes to anyone who filled out the puzzle correctly. When a slew of people approached us, waving completed puzzles in their hands, we knew we'd made a mistake. If you're going to lay down the gauntlet at Google, make sure you cap the prizes . . . you don't want to be bankrupted by a generous giveaway.

Surround Yourself with Wonderful People

For all the joking, this is the real secret to success. My business partner Will Pearson astounds me with his talents every day. Our company lucked into the kindest investors and some of the best mentors two 20-year-old kids could have ever asked for. And the amount of talent that's come through *mental_floss* is remarkable. Just scan the contributors to this book, and you'll find bestselling authors, editors at top blogs and magazines, and a few names of people who aren't famous yet, but soon will be. And the talent we've managed to hold on to is pretty extraordinary as well. I'd like to thank Jason English for curating such a rich and addictive website, Winslow Taft for always performing under such horrible deadlines (my fault!), Mike Rogalski for his lightning-speed layout, and Ethan Trex, who did the hero's work in assembling the content of this book. You make my job easy.

Happy reading,
Mangesh Hattikudur

10 CHEAT SHEETS FOR IMPRESSING A DIPLOMAT, PRESIDENT, OR POPE

HAIL TO THE POORLY DRESSED CHIEF!

FIVE PRESIDENTIAL FASHION FLUBS

1. Thomas Jefferson sometimes greeted dignitaries while wearing his PJs. On one such occasion, British minister to the United States Andrew Merry was on the receiving end of Jefferson's casual attire. He was not happy about it, writing,

   "I, in my official costume, found myself at the hour of reception he had himself appointed, introduced to a man as president of the United States, not merely in an undress, but ACTUALLY STANDING IN SLIPPERS DOWN TO THE HEELS, and both pantaloons, coat and under-clothes indicative of utter slovenliness and indifference to appearances, and in a state of negligence actually studied."

2. Although the Revolutionary War was long over by the time James Monroe took his post, he insisted on dressing as if the war was still raging outside the White House. That means britches, a buffcoat, a powdered wig, and a cocked hat. It was outdated and a little bit odd.

3. Zachary Taylor was so unconcerned about his appearance that he wore clothes and hats that were battered beyond belief. They were so worn and abused that it wasn't uncommon for people to see him out and about and mistake him for a farmer.

4. **Technically, this incident happened before Eisenhower was president, but it seems so out of character for the persona Ike had in public that we had to share it. Our 34th president had a wild sense of humor while he was attending West Point—once when his commanding officer requested that he appear in his dress coat, Eisenhower complied. He showed up to the meeting wearing his dress coat and not a stitch of any other clothing.**

5. Richard Nixon's fashion faux pas wasn't because of something he wore himself—it was because of the "uniforms" he had made for the White House police force. Inspired by the men that guard Buckingham palace, Nixon ordered formal garb complete with epaulets, embroidery and tall, fuzzy caps. After he was roundly ridiculed for the ensembles, which looked more like marching band uniforms than official police gear, he donated them to—what else—a marching band in Iowa.

A portrait of President Reagan made from 10,000 Jelly Belly beans hangs in the Reagan Presidential Library.

GIVE THEM LIBERTY

FIVE MODERN INDEPENDENCE MOVEMENTS

You've heard all about Palestine and Tibet, Quebec and Chechnya. But those aren't the only places that want to be sovereign. Here are five more would-be countries looking forward to paying U.N. dues.

Alaska's Grumpy Cold Men

For decades, a well-organized separatist movement has campaigned to turn America's largest state into its own nation. The bitterness dates back to 1958, when Alaska's citizens were given a simple yes-or-no vote on statehood. Many Alaskans felt they were denied more options on the issue, prompting a land developer named Joe Vogler to organize a revote that would offer Alaskans four possibilities—remain a territory, become a state, take commonwealth status, or become a separate nation.

George Washington didn't really have wooden teeth. He did, however, have a set of dentures made of hippopotamus bone.

Using the vote as his platform, Vogler ran for governor in 1974—and soon made a habit of it. With colorful slogans such as, "I'm an Alaskan, not an American. I've got no use for America or her damned institutions," Vogler spearheaded the Alaskan Independence Party (AIP), and his campaign has twice topped five percent of the vote. More surprisingly, former U.S. interior secretary Wally Hickel got elected governor on the AIP ticket in 1990. Unfortunately for the party, Hickel only ran on the ticket because he lost the Republican primary. Never a supporter of the plebiscite idea, Hickel left the AIP and rejoined the Republicans in 1994.

Today, the AIP continues to draw about four percent of voters statewide. And in 2006, Alaska took part in the first-ever North American Secessionist Convention, joining other groups from Vermont, Hawaii, and the South. As for Vogler, he was murdered in 1993—reportedly the result of an argument over a business deal. On a brighter note, honoring his wish to never be buried in U.S. soil, Vogler was laid to rest in Canada's Yukon Territory.

2 Vermont: Not-So Syrupy Sweet

Alaska isn't the only state that yearns to break away. In Vermont, a group called The Second Vermont Republic wants the state to return to independence. After all, Vermont was a republic from 1777 to 1791, when it became the 14th state in the nation.

The guiding principles of the Second Vermont Republic are generally progressive, with a focus on equality, green energy, sustainable agriculture, and strong local government. While most people in Vermont endorse these values, secession has been a tough sell. Still, the state independence movement is gaining ground, and one poll estimates that 13 percent of the populace supports the idea.

3 Gibraltar: Between a Rock and a Hard Place

Great Britain officially acquired Gibraltar from Spain in the Treaty of Utrecht in 1713, and Spain has been trying to get it back ever since.

The truth is, Britain would love to grant independence to the 2.5-square-mile island, but there's a catch. According to the treaty, Spain gets the territory should Britain ever relinquish it. And the people of Gibraltar don't want that. In 1967, Gibraltar's citizens voted on which country they'd rather belong to. With a 96 percent voter turnout, they favored Britain over Spain 12,138 to 44. Of course, Spain didn't take kindly to the decision and closed its border with Gibraltar, cutting it off from Europe by land for 16 years.

In 1973, Mao Zedong told Henry Kissinger that China had an excess of females and offered the United States 10 million Chinese women.

More recently, talks between Spain, Britain, and Gibraltar produced a 2006 agreement in which Spain agreed to ease its customs process and restrictions on air traffic. And in 2007, a new constitution gave Gibraltar greater autonomy under the crown, setting aside the Utrecht fight for another day.

4 Sealand: One Man Is an Island

If the existence of Sealand proves anything, it's that one country's trash can be another man's treasure. After World War II, Great Britain abandoned a series of military bases off its eastern coast. Seeing potential in one of the empty forts, former Major Roy Bates decided to claim it for his family. Then in 1966, he dubbed the island Sealand and declared independence. The following year, he fired warning shots at British naval vessels that dared to breach his waters.

When the British government brought Bates to court following the incident, they found they couldn't arrest him. Sealand was in international waters, just far enough off the coast to fall outside of British jurisdiction, so the island effectively got its sovereignty. But that was hardly the last time Bates had to fight for Sealand. In 1978, while Bates was abroad in Britain, a group of Dutch businessmen came to the island to supposedly discuss a deal. Instead, they kidnapped Bates' son and captured the fort. Naturally, Bates returned with a small army, fought the invaders, imprisoned them, and negotiated their release with their home country.

5 Greenland: Saving Up for Independence

Like a recent college graduate, Greenland wants to be on its own but just can't afford it yet. Denmark took control of the ice-capped landmass in 1721 and has been gently nudging it out the door for decades. In 1953, the Danes upgraded Greenland from a colony to an overseas county and gave it representation in parliament. And in 1979, they backed off even further, handling little more than Greenland's foreign policy and defense. Yet, Denmark still pays about half of Greenland's domestic budget, at a cost of about $650 million annually. Polls in Denmark show that the majority of the population supports the idea of letting Greenland's 57,000 inhabitants vote for independence. In other words, Greenland can be free if it wants.

Strangely, global warming may give Greenland the financial boost it needs to leave Denmark. As Arctic ice melts, the island's natural resources will become more accessible. The U.S. Geological Survey estimates that Greenland's northeast coast alone could produce more than 30 billion barrels of oil, and a few major oil companies have already bought permits to explore the land. The mining of gold, zinc, and other minerals is on the rise, too. Last year, aluminum giant Alcoa announced its intention to build the world's second-biggest smelter there. Plus, Greenland is investigating how to use the melting ice to expand its hydroelectric power industry. If it all adds up, Greenland may be moving away from the motherland sooner than it thought.

As a Congressional fellow, Dick Cheney worked for one day as Deputy Press Secretary for Senator Ted Kennedy.

WASHINGTON'S LEOPARD-SKIN ROBES AND THREE OTHER PRESIDENTIAL SECRETS LEFT OFF THE WHITE HOUSE TOUR

1 The Fabulous Life of George Washington

As president, ol' Georgie pulled in a salary of $25K a year. That's roughly $1 million in today's currency. Apparently excited by his newfound purchasing power, Washington started living it up, reportedly buying leopard-skin robes for all his horses and spending seven percent of his income on alcohol.

Tire mogul Harvey Firestone gave Calvin Coolidge a pygmy hippopotamus in 1927. Many of the pygmy hippos in U.S. zoos are its descendants.

John Quincy Adams' Naked Swimming Fetish

Forget secret tapes and shredded documents. Back in the early 19th century, there was a better way to get a glimpse of an American president truly exposed. All you had to do was show up at the banks of the Potomac River early in the morning during the warmer months between 1825 and 1829 to catch John Quincy Adams skinny-dipping.

3 Bush Leaves His Mark on Japan

If you remember one thing from the first Bush administration, it's probably not the 1992 state dinner during which President George H. W. Bush, ill with the flu, lost his lunch in the lap of the Japanese prime minister. Well, a lot of Japanese remember that incident a little better. Turns out, Bush's faux pas coined a Japanese slang word, *bushusuru*, which translates as "to do the Bush thing," meaning "to vomit."

4 Thomas Jefferson Goes Missing from Office

What do you want on your tombstone? Thomas Jefferson knew, so he took the time before he died to write out the inscription. A rather lengthy memorial, the missive listed Jefferson's many great accomplishments, from "author of the Declaration of Independence" to "founder of the University of Virginia." However, he did forget one small achievement. The tombstone fails to mention that Jefferson was once president of the United States.

Washington Wasn't the First American President.

Your teachers all said G.W. was the first American president, but George "I Cannot Tell a Lie" Washington would have told you differently. During the American Revolution in 1781, the Continental Congress elected Maryland statesman John Hanson to the post of President of the United States in Congress Assembled. After Washington defeated the British at the Battle of Yorktown, Hanson sent him a congratulatory note. Washington's reply was addressed to the "President of the United States." Not until he was elected in 1789 did Washington officially take his own version of the title.

Andrew Johnson was buried with his body wrapped in an American flag and his head resting on a copy of the Constitution.

THE VERY FEW, THE PROUD
THREE OFFBEAT MILITARY UNITS

1 High and Dry: The Bolivian Navy

Bolivia boasts a navy that's 5,000 sailors strong and has the boats to match. It lacks a crucial ingredient for true naval dominance, though: The country doesn't feature a single foot of coastline.

The concept of a Bolivian navy used to make a bit more sense. The country originally included a coastal region, but after Chile won the War of the Pacific in 1884 it annexed the area that included Bolivia's beaches. Over 125 years later, Bolivians are still peeved about the loss. In fact, they're so irritated that leaders re-formed the Bolivian Navy in 1963 to help maintain the country's true identity as a coastal nation.

Luckily for Bolivian taxpayers, though, their navy has practical value beyond thumbing the country's collective nose at a 19th-century conflict. While Bolivia doesn't have a coast, Amazonian rivers crisscross the country, providing drug smugglers with perfect routes for moving merchandise. The Bolivian Navy plays a key role in curbing this sort of illegal activity in addition to patrolling Lake Titicaca, which straddles the Bolivia-Peru border in the Andes.

Bolivian leaders are doggedly attempting to coax Chile into returning at least a little sliver of the coast the country misses so much, and although change doesn't seem imminent, it's conceivable that a Bolivian beach might one day be more than an oxymoron. Until then, though, Bolivians have their navy to remind them that while maps may show their country is landlocked, its citizens know better.

2 Efficient as Clockwork: The Swiss Army

Switzerland may be famed for its neutrality, but don't mistake it for a country full of pacifists. As part of its armed neutrality Switzerland puts nearly every able-bodied Swiss man through military training, and they're ready to fight with more than just red pocketknives. As a Swiss officer once told Pulitzer Prize winner John McPhee, "Switzerland does not have an army. Switzerland is an army."

The Swiss Army is more of a citizen militia than a traditional standing army. Universal conscription forces every male into service at age 20. After these conscripts receive their training they're free to start their careers, but they'll remain in the military's reserves for much of their adult lives and receive periodic recalls for training. In keeping with the citizen-militia concept, soldiers don't return their guns to an armory when their training is over; instead, they store their weapons in their homes. Swiss neutrality makes the citizen-soldiers' mission extremely straightforward: protecting Swiss safety and independence against outside attackers.

The Swiss Army first ordered its famous knives in 1889 after switching to a rifle that required a screwdriver for assembly.

To some Swiss, these defenses seem a bit unnecessary, and in the current world climate it does seem unlikely that an acquisitive power would just march into Switzerland only to be repelled by rifle-toting chocolatiers. (The Swiss Army hasn't actually clashed domestically with a foreign enemy since Napoleon pulled out of the country in 1815.) Switzerland's citizen-soldiers remain at the ready should anyone be foolish enough to enter their territory, though. Invaders beware: if this army has the necessary ingenuity to integrate a toothpick into a pocketknife, just think what sort of brilliance it could muster on a battlefield.

3 (Briefly) Up in the Air: The Republic of Singapore Air Force

The Republic of Singapore Air Force has funding and state-of-the-art equipment that would make many countries' airmen jealous. There's a hitch, though: Singapore doesn't have much airspace. The entire country is smaller than New York City, which means there's not much room for squadrons of jets and helicopters to fly around. Since pilots can't train in the airspace equivalent of a phone booth, Singapore has had to stash parts of its air force in friendly countries around the world.

While Singapore maintains several air bases on its own soil, aviation fans can also catch a glimpse of Singaporean aircraft on several other continents. Singaporean pilots train in such far-flung places as Luke Air Force Base in Arizona, Western Australia, India, and France's Cazaux Air Force Base.

Why do other countries welcome Singapore's forces in for training? Because Singapore isn't just the military equivalent of a freeloading friend who crashes on your couch. Given its geographic position in Southeast Asia and its largely ethnic Chinese population, Singapore is a valuable tactical and political ally for Western countries trying to gain a foothold in the area. Singapore also reciprocates the hospitable treatment it receives abroad; thanks to a number of agreements brokered since 1990, the United States has access to military facilities on Singapore's soil.

Plus, Singapore's pilots are downright gracious guests. When Hurricane Katrina struck New Orleans, Singapore quickly offered up its Texas-based detachment of Chinook helicopters and crews to help with the relief effort.

Nicorette was originally commissioned by the Swedish Navy in an attempt to get submariners to stop smoking on board.

ONE MORE INDEPENDENCE MOVEMENT (BECAUSE WE CAN'T STOP THINKING ABOUT THEM!)

1 The Conch Republic

In the early 1980s, the U.S. Border Patrol set up a checkpoint at the entrance to the Florida Keys in an effort to stop illegal drugs and immigration. Checking every single person's ID is a time-consuming process, though, and the checkpoint gave rise to a 20-mile traffic jam that turned tourists away. The downturn in visitors stifled the Keys' economy, so residents tried to get the checkpoint taken down. After numerous failed legal attempts to have the checkpoint removed, on April 23, 1982, Key West mayor Dennis Wardlow declared the Florida Keys were seceding from the Union.

Moments later, now-Prime Minister Wardlow symbolically declared war on the United States by breaking a stale piece of Cuban bread over the head of a man dressed in a U.S. Navy uniform. One minute later, Wardlow turned to the Admiral in charge of the U.S. Naval Base at Key West and surrendered, thus ending the Conch Republic's Civil Rebellion. He then immediately asked for $1 billion in federal aid to help rebuild his war-torn nation's economy.

While the Republic only officially existed for one minute, the tongue-in-cheek spirit of the rebellion lives on. Today you can buy Conch Republic citizen and diplomatic passports (both of which have reportedly been used for international travel, though they are not official legal documents) and even an official flag of the republic complete with the awesome motto "We seceded where others failed." The community has even minted a series of limited edition one-conch dollar coins that can be used as "legal tender" while in the Keys.

James Buchanan was nearsighted in one eye and farsighted in the other. To see correctly, he tilted his head at an odd angle.

EIGHT FLAG FACTS FOR THE NEXT STATE DINNER YOU HOST

1. Libyan dictator Muammar Gaddafi designed his own flag for the country, and nobody's going to confuse him with Betsy Ross. The Gadaffi-made flag is simply a plain green rectangle, the world's only national flag that features only one solid color and no design.

2. Mozambique takes the title of World's Most Violent Flag. It's hard not to win that competition when your flag's design includes an AK-47. The well-armed flag was originally the banner of the Mozambique Liberation Front that fought against colonial rule. When the country achieved independence, the Mozambique Liberation Front morphed into the dominant Frelimo Party, and the banner became the national flag.

A vexillologist is a person who engages in the scholarly study of flags.

3 As American troops stationed in Afghanistan learned in 2007, it pays to be very, very careful about using the Saudi Arabian flag on commercial items. The United States figured this out the hard way after giving Afghan children soccer balls with different countries' flags placed on every panel.

Sounds like a nice gesture, right? It would have been if the ball hadn't included the Saudi flag. The green flag features white Arabic writing of the shahada, the sacred Muslim declaration of faith that is one of the Five Pillars of Islam. Putting such an important component of Islam—including the name of Allah—on a soccer ball was incredibly offensive to the Afghan people, and the American military had to quickly apologize for its inadvertent insensitivity.

4 Nepal boasts the world's only national flag that's not rectangular or square. Instead, the flag looks like two triangular pennants stacked on top of each other. Originally these two pennants flew separately, but modifications later joined them into a single offbeat shape.

5 Hosting a diplomat from Chad but realize you don't have a Chadian flag handy? Dust off your Romanian one. Both flags consists of nearly identical vertical bars of blue, yellow, and red. Romania uses a slightly lighter shade of blue, but you could probably pull off a switcheroo without causing an international incident.

6 In most countries, the national flag should only be flown upside down as a sign of extreme danger or dire circumstances. Things are a bit different in the Philippines, though. When the nation goes to war it raises its normal flag upside down as its war flag.

7 It's hard to pin down the exact origins of the tradition of flying a flag at half-staff in times of mourning. The oldest commonly accepted reference to a half-staff flag dates back to 1612, when the captain of the British ship *Heart's Ease* died on a journey to Canada. When the ship returned to London, it was flying its flag at half-mast to honor the departed captain.

According to one line of scholarly thinking, by lowering their flag, the sailors were making room for the invisible flag of Death. This explanation jibes with the British tradition of flying a "half-staff" flag exactly one flag's width lower than its normal position to underscore that Death's flag is flapping above it.

8 Why does the United Nations' flag feature a blue background? Its designers selected blue as the opposite of red, the color of war.

According to Section 8 of the U.S. Flag Code, "The flag should never be used as wearing apparel, bedding, or drapery."

THAT'LL LOOK GREAT ON A BUMPER STICKER

THE 10 GREATEST CAMPAIGN SLOGANS IN PRESIDENTIAL HISTORY

❶ Voters didn't know much about Democrat Franklin Pierce when he headed into the 1852 election, so Pierce decided to cast himself as the rightful heir to popular ex-president James K. Polk. Pierce's pun of a slogan? "We Polked You in '44, We Shall Pierce You in '52." It may sound oddly threatening now, but it did the trick. Pierce beat his Whig opponent in a landslide.

❷ Modern-day politicians make some pretty outlandish campaign pledges, but giving away government property has to take the cake. That's what Abraham Lincoln did in 1860 when he ran for the White House under the slogan "Vote Yourself a Farm"—a bold promise to give settlers free land throughout the West. To his credit, however, Lincoln followed through and signed the Homestead Act in 1862.

During the 1828 campaign, Andrew Jackson's supporters claimed John Quincy Adams had offered his wife's maid to Russian Czar Alexander I.

Modern politicians didn't invent you're-either-with-us-or-against-us politics. Way back in 1868, General Ulysses S. Grant rode his Civil War victories into the White House with the slogan "Vote as You Shot"—a direct order to Union voters to toe the Republican line.

4

The award for quickest about-face on a campaign slogan goes to Woodrow Wilson, who campaigned for re-election in 1916 with the motto "He Kept Us Out of War." Americans voted with him in an effort to keep the peace, but five months later, Wilson led the country into World War I.

Prohibition was all the rage in 1920, much to the dismay of Democratic nominee James M. Cox, who believed making alcohol illegal only benefited criminals and bootleggers. His opponent, Warren G. Harding, attacked Cox for this stance and ridiculed him with the slogan "Cox and Cocktails." Ironically, after Harding won the presidency in a landslide, he was known to enjoy stiff drinks in the comfort of the White House.

6

Kansas Governor Alfred Landon emphasized his heartland roots during the 1936 election by adorning his campaign paraphernalia with bright yellow sunflowers. In response, opponent Franklin Roosevelt and his Democratic supporters went right for the kill, pointing out that "Sunflowers Die in November." They were right; Landon won just two states. Kansas wasn't one of them.

7

When FDR sought an unprecedented third term during the 1940 presidential race, it incited a backlash among those who felt it was time to move on. His Republican opponent, Wendell Willkie, got right to the point, stamping his campaign buttons with the slogan "Roosevelt for Ex-President."

8

Republican nominee Barry Goldwater inspired a legion of impassioned conservatives in 1964 with his slogan "In Your Heart, You Know He's Right." But Lyndon Johnson's Democratic campaign came up with a response that more effectively branded Goldwater as a right-wing extremist: "In Your Guts, You Know He's Nuts."

9

After unexpectedly winning the 1976 Democratic presidential nomination, Georgia Governor Jimmy Carter sought to stress his humble roots as a peanut farmer and also prove that he was a candidate to take seriously. He did both with his slogan, "Not Just Peanuts."

10

Barack Obama may have shown up at the right time with his "Change" campaign, but he's not the first Democrat to try this approach. The party's 1984 nominee, Walter Mondale, campaigned with the slogan "America Needs A Change." Unfortunately for Mondale, America disagreed, and 49 states voted for incumbent Ronald Reagan.

During the bitter Adams-Jefferson election of 1800, Martha Washington called Jefferson "one of the most detestable of mankind."

BUTT OUT OR DIE

FIVE HISTORICAL BANS ON SMOKING

1 The Sultan Puts Out Smokers

When Sultan Murad IV took over the Ottoman Empire in 1623, he inherited a land filled with corruption and decadence. He made quick work of that hedonism, though, and by 1633 Murad had banned all tobacco, wine, and coffee from his empire. Murad IV made Pope Urban VII look like a pushover, too; his punishment for breaking the ban was death.

Murad IV didn't leave enforcement to his minions, either. He supposedly walked the streets of Istanbul in plain clothes and used his mace to execute anyone he caught using tobacco. As many as 18 people a day met their demise for smoking until Murad's successor, Ibrahim the Mad, lifted the ban.

At around the same time, Russia instituted a similar ban. First-time offenders would get a slit nose, take a beating, or be exiled in Siberia. Repeat offenders earned themselves an execution. These stiff penalties hung around until Peter the Great came to power in 1682.

2 The Pope Cracks Down on Smoke

Pope Urban VII's papacy began on September 15, 1590. It ended with his death from malaria less than two weeks later. Although he didn't spend much time as the head of the Catholic Church, Urban VII was around long enough to make his feelings on tobacco known. He banned all tobacco "in the porchway of or inside a church, whether it be by chewing it, smoking it with a pipe or sniffing it in powdered form through the nose." The penalty for breaking his edict? Excommunication.

Scholars consider Urban VII's crackdown to be history's first public smoking ban. Various papal bans on smoking stuck around until 1724, when tobacco-loving Pope Benedict XIII gave Catholics the thumbs-up to light up again.

In 1943 Philip Morris ran an ad acknowledging "smokers' cough." They claimed it was caused by smoking brands other than Philip Morris.

King James' Ideal Version of England Is Smoke-Free

King James I of England was no fan of tobacco, but instead of whining about it, he picked up his pen. In 1604, James wrote the treatise *A Counterblaste to Tobacco*, and true to form for early 17th century pamphlets, the King didn't pull any punches, writing, "What honour or policie can moove us to imitate the barbarous and beastly maners of the wilde, godlesse, and slavish Indians, especially in so vile and stinking a custom?"

Ouch. Anti-Indian racism aside, James also warned of potential dangers from second-hand smoke and lung damage in addition to making a much simpler argument against tobacco smoke: it stinks. Later he refers to smoking as "a custome lothsome to the eye, hatefull to the Nose, harmefull to the braine, dangerous to the Lungs, and in the black and stinking fume thereof, neerest resembling the horrible Stigian smoke of the pit that is bottomlesse."

For someone with such strong feelings about smoke, James I surprisingly didn't ban tobacco altogether. He did, however, jack up excise taxes and tariffs on the weed by upwards of 4,000 percent. Interestingly, early 20th century tobacconist and writer Alfred Dunhill speculated in *The Pipe Book* that James' hatred of tobacco may have stemmed from how much the monarch loathed Sir Walter Raleigh, who was often seen smoking a pipe and actually turned Queen Elizabeth I on to smoking in 1600.

4 French Smokers Head to the Doctor . . . for More Smokes

French tobacco enthusiasts faced a bit of a curveball in 1635. They could still smoke, but they would have to buy their tobacco from an apothecary. They would also need a doctor's prescription. Luckily for smokers, this restriction didn't last too long. In 1637, King Louis XIII—the same King Louis XIII who enacted the laws in the first place—became a snuff man and repealed all of the anti-tobacco laws.

Colonists Turn on Their Cash Crop

Early American colonists may have made some nice loot selling tobacco, but that doesn't mean they were totally in favor of using it. In 1632, Massachusetts became wary of the fire danger from smoldering butts, so it banned public smoking. Connecticut followed suit in 1647 when it dictated that citizens could only smoke once a day, and even then one couldn't be a social smoker since the law dictated that smokers could only burn one when "not in company with any other." In the 1680s, Philadelphia joined in with a ban on smoking in the city's streets.

After 2000, Monopoly spokesman, "Rich Uncle Pennybags," changed his name to "Mr. Monopoly" and quit smoking.

SMALL NATION, BIG STORIES

FIVE SECRETS OF THE VATICAN EXPOSED

Vatican City may have fewer than 1,000 citizens and span only 110 acres, but it also has a multimillion-dollar budget and an unbelievably complex history. Understanding how it all works requires parsing through centuries of religious texts. Is the Vatican confusing and mysterious? Is the Pope Catholic?

1 Regular Exorcise!

Baudelaire once said that "the greatest trick the devil ever pulled was convincing the world he doesn't exist." But in modern-day Vatican City, the devil is considered alive and well. The former Pope John Paul II personally performed three exorcisms during his reign, and the current Pope Benedict XVI is expanding the ranks of Catholic-sponsored exorcists throughout the world. In fact, Father Gabriele Amorth, the Church's chief exorcist, claims to expel more than 300 demons a year from the confines of his Vatican office, and there are more than 350 exorcists operating on behalf of the Catholic Church in Italy alone. Amorth also teaches bishops how to tell the difference between satanic possession and psychiatric illness, noting that those who suffer from the former seem to be particularly repulsed by the sight of holy water and the cross.

2 Where Thieves Go to Prey

With 1.5 crimes per citizen, Vatican City has the highest crime rate in the world. It's not that the cardinals are donning masks and repeatedly robbing the bank, it's just that the massive crowds of tourists make Vatican City a pickpocket's paradise. The situation is complicated by the fact that the Vatican has no working prison and only one judge. So most criminals are simply marched across the border into Italy, as part of a pact between the two countries. (The Vatican's legal code is based on Italy's, with some modifications regarding abortion and divorce.) Crimes that the Vatican sees fit to try itself—mainly shoplifting in its duty-free stores—are usually punished by temporarily revoking the troublemaker's access to those areas. But not every crime involves theft. In 2007, the Vatican issued its first drug conviction after an employee was found with a few ounces of cocaine in his desk.

The Vatican Bank is the world's only bank that allows ATM users to perform transactions in Latin.

They Have the Finest Swiss Bodyguards

Nowadays, the Swiss have a reputation for pacifism, but back in the 1500s, they were considered an unstoppable military force. Swiss armies were renowned for their mastery of a weapon called the halberd, a deadly combination of a spear and an ax, and their ground troops were famous for routinely demolishing legions of enemies on horseback. After Pope Julius II witnessed their ferocity in battle 500 years ago, he recruited a few soldiers to become his personal bodyguards. Ever since, Swiss Guards have pledged fidelity to the Pope, sometimes dying for the cause. During the sacking of Rome in 1527, for instance, three quarters of them were killed while providing cover for Pope Clement VII to escape.

Today, the hundred or so members of the Swiss Guard spend most of their time bedecked in Renaissance garb, twirling their halberds in ceremonies or manning checkpoints around the Vatican. When the Guards are actually protecting the Pope, they wear plain clothes and carry distinctly modern weapons.

Pope John Paul II once urged the media to stop using the term "Popemobile" because he found it undignified.

4 You Can Read the Pope's Mail

The Vatican's secret archives haven't been truly secret since Pope Leo XIII first allowed scholars to visit in 1881. Today, it's even more accessible. Outsiders are free to examine the correspondences of every pope for the past 1,000 years, although there is one catch: Guests have to know exactly what they're looking for. With 52 miles of shelves in the archives, the librarians prohibit browsing.

The most famous letter there is probably Henry VIII's request that his marriage to Catherine of Aragon be annulled, which Pope Clement VII denied. Henry divorced Catherine anyway and married Anne Boleyn (and four other women), leading to Rome's break with the Church of England. The archives also contain an abundance of red ribbons, which were used to bind 85 petitions from English clergymen and aristocrats.

5 Faith-Based Economics

The Vatican needs several hundred million dollars per year to operate. Its many financial responsibilities include running international embassies, paying for the Pope's travels around the world, maintaining ancient cathedrals, and donating considerable resources to schools, churches, and health care centers. So where does that money come from? Catholics pay tithes to their local parishes and donate about $100 million every year to the Vatican itself. But collection plates aren't the Vatican's only source of money. The city-state also gets cash from books, museums, stamps, and souvenir shops. (Get your limited-edition Vatican euros here!)

But that's not always enough. By the end of 2007, the city-state was $13.5 million in the hole. Part of the problem was the weakened American dollar, which translated into less purchasing power. Another contributing factor was the lackluster performance of the Vatican's newspaper, *L'Osservatore Romano*. To boost subscriptions, the Pope has asked the editor to spice up the layout with more photos and allowed him to cover world news stories in addition to the traditional religious fare.

Pope Leo X, a notoriously free spender, once had to pawn his own palace furniture and silver to cover his luxurious lifestyle and patronage of the arts.

GOVERNMENTS FUND THE DARNEDEST THINGS

FOUR OF THE LARGEST, ODDEST, AND MOST USELESS STATE PROJECTS IN THE WORLD

1 Dumb as a Limestone Brick: Indiana's Misguided Bid for Tourists

THE GREAT IDEA: Turn a small Midwestern town into a tourist mecca for lovers of limestone block.

THE GREAT BIG PROBLEM: Limestone block is not as big a draw as you might think.

COST TO TAXPAYERS: $700,000

Despite being the undisputed "Limestone Capital of the World," Bedford, Ind., always had a hard time figuring out how to parlay its claim to fame into a thriving tourism industry. That is, until Bedford Chamber of Commerce member Merle Edington came up with a brilliant plan. In the late 1970s, Edington proposed that Bedford build a Disney-style theme park.

But, instead of cartoon characters, the park's main attraction would be limestone, featuring a 95-foot-high replica of the Great Pyramid of Cheops built out of (you guessed it) local limestone blocks. And, on the off chance that a scale model of one world wonder wouldn't be exciting enough, Edington added plans for an 800-foot-long replica of the Great Wall of China. While the power of limestone over the vacationing public is debatable,

The Christmas classic "Do You Hear What I Hear?" was written in 1962 as a plea for peace during the Cuban Missile Crisis.

Edington convinced the Commerce Department's Economic Development Administration to believe in his dream—to the tune of $700,000.

Unfortunately, those funds dried up quickly, thanks to Wisconsin senator William Proxmire (famous for his "Golden Fleece Awards" ridiculing government waste), who called attention to the project. The town was left deep in debt, unable to even pay Edington's salary. Today, the abandoned project is little more than a giant rock pile.

2 Building a Better Future, One Spokesmodel at a Time: Vanna White School Hits Italy

THE GREAT IDEA: When creating jobs for the unemployed, cater to the pretty people first.
THE GREAT BIG PROBLEM: Regional dignity seems to have gone overlooked.
COST TO TAXPAYERS: 1 million euros (about $1.2 million U.S.)

In 2003, the European Union came up with a novel solution for lowering soaring unemployment levels among young people in Italy's Campania region—history's first job-creation program catering solely to the physically attractive. With a grant of 1 million euros, the EU opened First Tel School, a quasi-educational program designed to train students to become TV game show hostesses. The eight-month program, labeled "bimbo school" by critics, offered courses on diction, makeup, and other skills necessary to get a job on one of the country's many popular game and variety shows.

The school's de facto policy of discriminating against all of Campania's less toothsome residents aside, First Tel also failed to admit enough students to offer any kind of relief to the region's unemployed population (which was estimated to be 50 percent of young people). Fewer than 100 spots were made available to the 1,200 Vanna White wannabes who applied. Worse still, even those who did make the cut received a whopping 2 euros for every class attended, as well as auditions for parts in the shows First Tel produced for Italian television stations. Meanwhile, rejected applicants remained unemployed and got daily, televised reminders of how unappealing they were.

3 The Mess with Texas: Superconducting That's Less than Super

THE GREAT IDEA: Build a miracle machine that can replicate the Big Bang, help treat life-threatening illnesses, and maybe even unfold the mysteries of the universe.
THE GREAT BIG PROBLEM: You get what you pay for—and miracle machines cost way, way too much.
COST TO TAXPAYERS: Roughly $12 billion—and the lives of billions of innocent atoms.

Few government projects have ever been announced with the level of fanfare reserved for the 1980s Superconducting Super Collider. Housed in a 54-mile underground tunnel beneath Waxahachie, Texas, the Super Collider was designed to accelerate beams of subatomic particles to fantastic speeds and then crash the particles into one another, purportedly generating

Uncle Sam was based on Samuel Wilson, who worked as a meat inspector during the War of 1812.

huge amounts of energy. Advocates believed the machine would be able to simulate the conditions present during the Big Bang, thus allowing scientists to gain new insights into the very nature of matter.

But many Super Collider fans made even bolder statements about the machine's capabilities, pointing out that other devices using similar technology had been used to treat cancer and learn more about HIV. As potential uses for the machine grew, however, so did the cost—ballooning from an original estimate of less than $5 billion to just under $12 billion.

Finally, in the midst of the 1993 budget-cutting boom, Congress pulled the plug on the project, with less than one-third of the tunnel finished. For a while, it was used to store Styrofoam cups, but then it was sold off to private businesses for pennies on the dollar. Although scientists (and the citizens of Waxahachie) still mourn the loss of this major research center, there are several other machines in the world that do basically the same thing on a smaller scale. They're called particle accelerators, and the largest one is a mere 5 miles in diameter.

4 A Cure for The Cure: Combatting Teen Angst in Missouri

THE GREAT IDEA: Cure adolescent angst by fighting the forces of goth subculture in Missouri.

THE GREAT BIG PROBLEM: Harassment of misunderstood youth subcultures is best left to high school football players. They'll do it for free.

COST TO TAXPAYERS: $141,000

In 2002, Blue Springs, Mo., found itself under siege from an insidious force—a force armed with pale skin, black hair, black eyeliner, and a collection of old albums by The Cure. Yes, goth culture had gained a toehold in the otherwise ordinary hamlet, and there was only one thing to do about it: Stamp it out with the aid of a $273,000 grant from the U.S. Department of Education.

With the funds secured, the Blue Springs Youth Outreach Unit set out to combat the goths, whom the community believed to be involved in activities including animal sacrifice, self-mutilation, and vampire worship. Reportedly, about half of the grant went to staff salaries, staff trips (to conferences on teen drug use and satanic cults), the formation of a little-used counseling program, and a series of never-held town-hall meetings (presumably intended for daylight hours, when vampire infestation wouldn't be a problem). After two years of public criticism, officials made the startling discovery that maybe goths weren't such a problem after all, and returned the remaining $132,000 to the government.

Alaska is so big you could fit 75 New Jerseys in it.

REAL U.N., LESS-THAN-MODEL CITIZENS

SEVEN SHAMELESS ABUSES OF DIPLOMATIC IMMUNITY

1 Park Wherever You Want

The most common manifestation of diplomats' inconsiderate behavior involves their parking habits. Just ask New Yorkers; diplomats at the United Nations apparently view Manhattan as their private parking lot. In 1996 alone, diplomats racked up 143,508 parking summonses, which would have cost them $15.8 million if not for diplomatic immunity. Russia was responsible for nearly 32,000 of those fines.

2 Or Clear a Spot for Yourself

A few hundred thousand unpaid tickets look like downright responsible behavior when compared to former Afghani diplomat Shah Mohammad Dost's antics behind the wheel. In 1987, Dost was accused of intentionally running a woman over in order to get a parking spot during a trip to an appliance store in Queens. According to the victim, her boyfriend was backing into the spot when Dost rolled up and demanded they cede the space to him because he was a diplomat. When they refused, Dost threw his Lincoln into gear and ran the woman over, sending her to the hospital.

3 Become a Doggie Diplomat

Here's a case where diplomatic immunity didn't work out quite as well as a diplomat had hoped. In 1975, a U.N. delegate from Barbados claimed that diplomatic immunity extended to his pooch, who had bitten several people. The delegate warned police officers of "possible international consequences" if they tried to contain the aggressive German shepherd. Nice try, but Fido's not exactly negotiating trade treaties.

A Mexican diplomat got the same rude awakening in 1984. Military attaché Enrique Flores was keeping a pack of 10 basset hounds at his Virginia home in violation of local zoning laws. Even though the laws stated Flores could only have four hounds at once, he appealed to the State Department for diplomatic immunity. The State Department turned him down. Guess they're cat people.

4 Lose Your Immunity in the Divorce

In 1989, Mozambique's representative to the United Nations wanted to divorce his American wife, so he waived his diplomatic immunity in order to take the matter to court. Unfortunately for the diplomat, Antonio Fernandez, he didn't fare well in the case; he ended up losing the couple's $5 million estate in the decision. Whoops.

Ronald Reagan once said, "The day after I was elected, I had my high school grades classified Top Secret."

Fernandez didn't suffer from any shortage of gall, though. After losing the decision he attempted to invoke his diplomatic immunity privileges to keep from paying his ex-wife. Fernandez took his case all the way to the Supreme Court, but in the end his former love got the couple's Greenwich, Conn., estate.

5 Light Up on a Plane

In 2010, a Qatari diplomat ran into trouble on a Washington-to-Denver flight when he decided to have a smoke in the plane's lavatory. To make things worse, Mohammed Al-Madadi also made a joke that some passengers and flight personnel perceived to be a terrorist threat. Air marshals sounded various alarms, and in the end two F-16 fighter jets escorted the flight to its final destination. While diplomatic immunity kept Al-Madadi from being charged with any crimes, the Qatari government sent him home to help smooth things over.

6 Stop Paying Your Rent

A word of advice to landlords out there: If diplomats want to rent one of your properties, you might want to get a hefty security deposit. Just ask some of Manhattan's biggest landlords. A 1996 *New York Times* story illustrated the difficulty of renting to diplomats; landlords really don't have any legal mechanism through which they can collect delinquent rent or evict diplomatic tenants. At the time the article was written, one West African country was over $20,000 behind in its rent for a pair of luxury apartments in midtown Manhattan.

If one of us normal folks pulled a stunt like that, we'd be out on the streets. Diplomats, on the other hand, enjoy a special kind of immunity known as "inviolability," which states that their private residences can't be entered by the host country's agents without the visiting country's consent. In short, the only way you can evict foreign diplomats is if their own country gives you the thumbs-up first.

7 Stop Cutting Your Lawn

In 2008, the residents of New Rochelle, New York, found themselves in a common dilemma: One long-vacant house in the community had become a real eyesore. Weeds filled the yard, the paint had gone bad, and the property found itself in an ugly state of decay. New Rochelle was powerless to fix the problem, though: The dilapidated house had a sort of diplomatic immunity that enabled it to be that run-down.

Somalia owned the house, which it occasionally used to house United Nations diplomats. Since the vacant house was exempt from taxes, the town couldn't use liens or other penalties to force the Somalians to do a little landscaping. The lesson here: If you want to shirk your lawn-mowing duties, join the Foreign Service.

Former First Lady Bess Truman was so underwhelmed by D.C.'s cleaners that she shipped her laundry home to Kansas City.

10 LISTS THAT SHOULD COME WITH A LAB COAT

THE WHY FILES

EIGHT QUESTIONS YOU PROBABLY NEED ANSWERED IMMEDIATELY

Why Does Hawaii Have Interstate Highways?

While we'd like to believe Hawaii's Interstate system exists for the sole purpose of annoying George Carlin, the name is actually a misnomer. Not all Interstates physically go from one state to another; the name merely implies that the roads receive federal funding. The three Hawaii Interstates (H1, H2, and H3) became Interstates as part of The Dwight D. Eisenhower System of Interstate and National Defense Highways to protect the United States from a Soviet invasion by making it easier to get supplies from one military base to another.

2 Can a Pregnant Woman Drive in the Carpool Lane?

Expectant mothers, start your engines! In 1987, a pregnant California woman was ticketed for driving "by herself" in the carpool lane. Sure, the citation was only for $52, but she sued anyway, contending that her 5-month-old fetus constituted a second person. Lo and behold,

The metal band that holds the eraser on a pencil is called a ferrule.

the jury agreed with her, despite the prosecution's argument that women could then just stuff pillows up their dresses to drive "carpool" on California's freeways. But as it turns out, the California Highway Patrol took care of that concern, brushing off the case as a bunch of hooey. Verdict or not, officers said they would continue to ticket solo drivers, even if they claimed to be pregnant.

3 What Makes No. 2 Pencils So Darn Special?

Little. Yellow. Identical. The No. 2 is definitely No. 1 in the pencil market. It's a staple in schools and workplaces everywhere, and the required writing utensil for Scantron® tests across the globe. But is it really that great of a pencil? You bet your bippy. No. 2's use medium weight graphite, which makes them the ideal pencils for general writing. 18th-century French pencil maker Nicolas-Jacques Conté created the number system based on a pencil's hardness (the higher the number, the harder the graphite), and we've been using it ever since. But let's not forget the other numbers of pencils out there. No. 1's are made with soft graphite and tend to smudge, and are often used to record bowling scores. No. 3's and above indicate harder pencils that are most often used for drafting, when you need a sharp, strong point.

4 Why Can't You Tickle Yourself?

Much to the dismay of wacky masochists everywhere, the human brain is wired against self-tickling. Because the brain controls movement, it knows what your hand is going to do before you do it. Thus it anticipates the exact force, location, and speed of the tickle and uses that information to desensitize you to your own roving hands. So why do we have a tickle response anyway? Turns out, it's a defense reaction meant to alert our cave-dwelling ancestors to creepy crawlies that didn't know their place, and the uncontrollable laughing fit that goes along with it is actually a panic response. Even if you know someone else is about to go for your rib cage, it's hard to turn the response off because a) your brain can't anticipate exactly how and where they'll tickle you and b) knowing someone is about to tickle you is usually enough to keep those panic receptors open and ready to go.

5 Why Don't School Buses Have Seat Belts?

Seat belts have been mandatory in cars for more than 40 years, so why aren't school buses equipped the same way? According to the National Highway Traffic Safety Administration, it's because school buses don't need seat belts to be safe. The bulkiness of a bus makes it about seven times safer than a passenger car. In the event of a collision, a bus can easily absorb the force of impact. Plus, kids riding in buses are doubly protected because the seats are designed to cushion children almost like eggs in a carton. The accommodations might not provide much legroom for unruly 8-year-olds, but the high seatbacks and heavy padding work to form a protective cocoon around them. If Junior is thrown forward in a crash, he won't get far before the cushy seatback absorbs his momentum.

Most school buses have been painted National School Bus Chrome Yellow since a 1939 national conference recommended it as the shade of choice.

Of course, none of this will help if the bus flips over. But the chances of that are so slim that most state legislators don't think seat belts are worth the added expense. Still, some states would rather be safe than sorry. New York and California, for example, now require all new school buses to come equipped with lap-and-shoulder belts.

6 Why Can't I Use My Cell Phone on an Airplane?

The Federal Aviation Administration (FAA) bars the use of all transmitting devices in the off chance that transmissions could interfere with a plane's navigation and communications equipment and cause system malfunctions. It's true that these concerns are overblown, but the FAA likes to err on the side of caution. (Can you blame them?)

The real reason authorities don't want you flipping open your mobile phone has less to do with crashing your plane and more to do with crashing the cell phone network. The Federal Communications Commission has determined that mid-flight calls have a direct impact on cell phone service on the ground. That's because cell phones are primarily designed for callers who are firmly planted on land, communicating with a single, nearby tower. If you're speeding through the sky at 550 mph, your phone will connect with multiple towers and eat up valuable space on their circuits, wreaking havoc on service. A 2007 plan to lift the ban was strongly opposed by cell carriers for this reason. So, at least for a while, frequent flyers should recline their seatbacks and enjoy the last place on Earth that's free of cell phone chatter.

7 Why Do Soap Operas Look Different from Other TV Shows?

Lighting, for one. Soaps and other lower-budget shows look "off" because they're often evenly lit across the entire set to facilitate simultaneously shooting with more than one camera. This lighting/shooting method means the actors can move around and the lights don't have to be reset for every shot. This allows for fewer takes and costs less, but it also means more diffuse, less natural-looking lighting in the final product.

The filming medium and technique make up the other half of the equation. Soaps have been shot on various types of videotape to keep costs down, and compared to prime time shows shot on film, they can look a little flat. Shooting with videotape also gives lower resolutions, so soaps have always made heavy use of close-ups to compensate.

8 Why Do Battery Letters Skip from A to C? Was There Ever a B-Cell Battery?

Battery letter designations are based on the size of the battery: For common sizes, A is the smallest, and D is the largest. By the same logic, AA batteries are larger than AAA. Unfortunately for B batteries, it's not the size that counts. You never see B batteries around because they aren't very useful. The size never caught on in products made for consumers, so stores didn't carry them, and the cycle continued. They are sold, but only in Europe, where they're used primarily to power bicycle lamps.

Dropping two aspirin tablets into a dead car battery can revive it!

FORGET YOUR DECILITERS

FIVE UNITS OF MEASUREMENT WEIRDER THAN THE METRIC SYSTEM

The Smoot

A smoot is exactly 5 feet, 7 inches—the height of MIT freshman Oliver Smoot in 1958 when he was used to measure the length of the Harvard Bridge between Boston and Cambridge. Smoot's fraternity brothers determined the bridge was exactly 364.4 smoots long, plus one ear.

Since weightlessness causes the spine to expand, astronauts may measure a few inches taller in space than they do on Earth.

The Beard-Second

A beard-second is the average length a man's beard grows in 1 second. However, experts disagree on what that length is. Some say 10 nanometers; others, including Google calculator, say it's 5.

3 The Wheaton

It's no surprise that delightfully geeky actor Wil Wheaton was one of the first celebs to embrace Twitter, but he was also one of the first to attract a massive number of followers. When half a million people subscribed to his Tweets, that number was dubbed a Wheaton. Today, the actor has 1.75 million followers, but the Wheaton has remained 500,000—meaning that Wil Wheaton actually has about 3.5 Wheatons.

The Helen

Helen of Troy's magnificent mug is said to have launched 1,000 ships. But what if there's just one ship that needs help getting out of port? Well, then you need a milliHelen. According to David Lance Goines and his Helen system of measurement, a picoHelen is the unit of beauty that inspires men to "barbecue a couple of steaks and toss an inner tube in the pool," whereas a teraHelen has the potential to "launch the equivalent of 1,000 trillion Greek warships."

5 The Sheppey

A herd of sheep can be picturesque from a distance, but the closer you get, the dirtier and more matted the wool looks. Fortunately, writers Douglas Adams and John Lloyd, coauthors of the humorous dictionary *The Meaning of Liff*, have given us a way to measure that distance. A sheppey is how far you need to stay away from a group of sheep so that they resemble cute balls of fluff. One sheppey is equal to about 7/8 of a mile.

In the 4th century BCE, Alexander the Great encouraged his men to shave so enemies couldn't grab their beards during melees.

THE FINAL COMMERCIAL FRONTIER

EIGHT EVERYDAY ITEMS BROUGHT TO YOU BY NASA (AND WHY TANG ISN'T ONE OF THEM)

Cordless Tools

As long as NASA was going to the trouble of sending Apollo astronauts to the moon, it figured it might as well equip them with drills and ask them to dig up rock samples when they got there. But realizing that a 239,000-mile extension cord would be impractical, NASA teamed up with Black & Decker to develop tools that featured rechargeable batteries and special low-power consumption motors, which should make your DustBuster seem a lot more impressive.

2 Smoke Detectors

In the 1970s, NASA partnered with Honeywell Corp. to create a device that would detect smoke and toxic gases in Skylab, America's first space station. The result was the first ionization smoke detector, using a minute amount of the radioactive isotope Americium-241. This led to the 1979 introduction of inexpensive photoelectric detection devices, which go off when smoke (or sometimes a hot, steamy shower) blocks the light beam. To date, smoke detectors have saved countless lives here on Earth, but they're especially useful in space, where running outside to wait for the fire truck isn't an option.

Enriched Baby Food

NASA-sponsored research has also helped make major improvements to commercially available baby food, and we're not talking about freeze-dried strained peas. While testing the potential of algae as a food supply for long-duration space travel, a Maryland-based biosciences company discovered an algae additive that contains two fatty acids closely resembling those found in human breast milk. The company now uses it to make an enriched infant formula called Formulaid, thought to be essential for babies' visual and mental development.

Buzz Aldrin consumed the first alcohol in space when he took communion after landing on the moon in 1969.

New-Age Pavement

When you buy a new set of tires, the old ones have to go somewhere, right? Most of them end up in huge, flammable tire dumps, which may hold millions of old tires, each one containing about a quart of oil in the rubber. If a dump catches fire, however, it can burn with a thick, toxic smoke for weeks on end.

But today, old tires are being put to good use. NASA's experience in fuel-related cryogenics helped develop processes to freeze the tires to below -200°F so that they crumble, separating the rubber from other materials and producing what's called "crumb." This waste is recycled into several new products, including an ingredient used to pave highways, which means your new radial tires may someday be rolling over your old ones.

5

Those Cool Ear Thermometers

Any parent knows you don't take a baby's temperature by sticking a glass thermometer in his mouth, but inserting it the other way isn't much fun, either. And what about the incapacitated patient who can't even say "aaah"? The Diatek Corp. of California wanted a safer way to take a person's temperature, and who better to turn to than NASA's Jet Propulsion Laboratory, the place with over 30 years of experience using infrared sensors to remotely observe celestial bodies? Together, they developed a fast and accurate thermometer that, when its disposable probe cover (to prevent cross-infection) is inserted into the ear canal, detects infrared radiation from the eardrum and gives a digital readout in less than two seconds.

Alan Shephard misfired on two balls during his famous foray into lunar golfing during the *Apollo 14* mission. The balls are still sitting on the moon.

6 Protective Paint

What do the Statue of Liberty, a gigantic Buddha in Hong Kong, and the Golden Gate Bridge all have in common? They're protected by the American space program . . . sort of. In the late 1980s, NASA's Goddard Space Flight Center began a research program to develop coatings for the Kennedy Space Center in Florida to shield the launch structures from salt-air corrosion, rocket exhaust, and thermal stress. Applications of this material proved ideal for protecting structures like bridges, antenna towers, and the occasional big Buddha.

7 Scratch-Resistant Glasses

Thanks to NASA technology, plastic lenses for glasses last up to 10 times longer than they used to. That's because its Ames Research Center created a scratch-resistant (read: extremely hard) coating to protect equipment from getting beaten up by space debris. Later, the Foster Grant Corp. acquired the license for the coating method and used it in their plastic sunglasses, which matched the hardness of glass lenses, but were much lighter. Among other uses, it's now employed in most eyewear and industrial face shields.

8 Oh-So Comfy Sneaker Insoles

Can't run a five-minute mile? Don't blame your sneakers. If they're relatively new, they're probably giving you quite a bit of help already. In the 1970s, many shoe manufacturers began replacing their standard foam rubber insoles with a new, highly shock-absorbent material—one giant step for tennis shoes. The new kicks were padded with "viscoelastic" bubbles that conformed to your foot and then returned to their normal shape when you took the shoes off. Turns out, they got the idea (and the technology) from NASA, which had developed the material to better cushion astronauts during blastoff.

. . . and One Pop Culture Killer:

Despite popular theory, NASA did NOT invent Tang orange breakfast drink for the astronauts. It was introduced in 1957 by General Foods and was on grocery store shelves for years before NASA decided it worked well in space.

Apollo 8 astronauts used Silly Putty to keep their tools from floating around in zero gravity.

THE CHEMICALS THAT WILL POWER THE FUTURE

Modern chemistry started off as a pipe dream (and soon became an elaborate con) as Renaissance-era alchemists mixed elixirs in a futile attempt to turn other metals into gold. Since then, though, chemists have identified or created the compounds that fuel our industry, keep us healthy, and form the building blocks of our civilization. But they're just getting started.

Helium-3

Nuclear power is at a crossroads. After the earthquake- and tidal wave-caused meltdowns at Japan's Fukujima plant, several countries are reexamining their interest in a power source that can cause so much devastation. While that may mean reverting to older energy sources like coal, it could instead fast track the development of fusion power to take the place of its more destructive fission brother. Fusion power is safer and easier to control than fission, but the technology to harness that power is still decades away.

Producing a single gram of antimatter costs about $62.5 trillion. Then it takes 40 billionths of a second for it to disappear.

Assuming we can iron out the kinks, though, there's already a great candidate for the fuel for these reactors—the Helium-3 isotope. Because of its simple structure and lack of radioactivity, it's safe and easy to work with. It's estimated that just 100 tons of the stuff could power everything on Earth for a year.

The only problem? There's not even that much of it here on Earth. Produced by nuclear reactions on the sun and blown across the universe by solar wind, Helium-3 has settled everywhere across the galaxy except Earth as it's repelled by the magnetic core. But a close neighbor doesn't have the same issues. The Apollo lunar missions of the 1970s found that the moon's soil is thick with the stuff—NASA, as well as the space programs of Europe, India, China, and Russia are all hoping to mine the estimated millions of tons of Helium-3 that has settled there.

2 Supercritical CO_2

If you put carbon dioxide over enough heat and under enough pressure, it begins to act more like a liquid than a gas. Called supercritical CO_2, this new fluid does some amazing things. Depending on how you pressurize it, it can take on the characteristics of many industrial solvents like chloroform—only without the toxicity. It's already being used to decaffeinate coffee beans.

In the future, it will be used to turn garbage and organic industrial waste into energy and useful products like wax. At extreme temperatures and pressures, supercritical CO_2 is also much more thermally efficient than steam, which means that turbines and engines that used to be powered by the latter can be made smaller and produce far more energy.

3 Methane Hydrates

Natural gas seeps from cracks in the deep ocean floor, creating mini-oases of strange creatures like tubeworms that are attracted by the gas-eating microbes. Sometimes, the gas bubbles all the way up out of the ocean into the greenhouse layer of the atmosphere, where it combines with all of the other gases, man-made and otherwise, to slowly heat the planet.

Most of the time, though, it falls back to the benthic soil, where it gets captured in a lattice of ice crystals. The resulting formations, called methane hydrates, dissolve as they're brought to the surface, so no one's really figured out how to mine them economically yet. But someone eventually will; these deposits are simply too enormous to ignore. When someone cracks the problem, we'll have a vast new source of fuel. It's estimated that there's more energy trapped in those hydrates than in any other fossil fuels on Earth.

Negative-40 degrees F and negative-40 degrees C are the same temperature.

Polymethine Dyes

In the future, you're never going to have to waste your time waiting for anything to download. Thanks to a group of Georgia Tech scientists, your Internet speed is going to be 50 times faster. Even the fastest routers today rely on fiber-optic technology that has to switch the data back and forth from light to electricity as it heads to its destination. The chemists have figured out a way to skip the electricity part of the network by creating a molecule that can keep the whole process optical. Called polymethine dyes, these molecules change their physical appearance based solely on the type of light shining on them. Once the chemists can figure out how to put these dyes into a switch, an all-optical two-terabit per second network will be ours.

5 Palladium

Once known mainly for its role as the carbon-monoxide destroying catalyst in your car's catalytic converter, the precious metal palladium is expanding its repertoire at a rapid pace. That's because it forces together the synthesized, complicated carbon-based molecules that make up new drugs, plastics, and industrial chemicals; in other words, it's the mortar for the building blocks of the modern world.

Palladium's unique chemical properties allow scientists to combine all sorts of compounds to mimic unusual supermolecules previously only found in nature. One of those, discodermolide, has anti-cancer properties but used to be only found in a rare Caribbean marine sponge. Using palladium, scientists can now create mountains of it from scratch in the lab. The element is also being used to create everything from light-emitting diodes to fungicide.

6 Green Fluorescent Protein

The crystal jellyfish isn't big or dangerous or rare. If you've ever hit the beach in California, you probably swam right by one. But this jelly is changing the entire field of biochemistry. That's because it possesses something called the Green Fluorescent Protein, or GFP, which glows under ultraviolet light. GFP is flexible, rugged, and nontoxic, meaning that biochemists can attach it to living proteins by splicing GFP genes into the genes that produce them. Initially, this was used as a novelty; you can still buy glow-in-the-dark mice from a company called NeonPets. But as scientists are learning how to manipulate GFP and follow these proteins in real time, our understanding of everything from how brain circuitry works to how viruses evolve has been redefined.

The infinity sign is formally known as a "lemniscate."

MIST OPPORTUNITIES

FIVE WEATHER EVENTS WORTH CHATTING ABOUT

Raining Rainbows

We've all heard about the damaging effects of acid rain, but what about colored rain? Over the course of an entire month in 2001, deep red rain fell in the Kerala region of India. Yellow, green, and black rain was also reported. The rain was such a deep color, residents claimed it stained clothes and resembled blood. The official report found that the unusual rain was caused by spores of a lichen-forming algae sucked into the atmosphere by a waterspout, much to the dismay of many people who thought it was caused by extraterrestrial activity.

Siberia experienced a strange yellow-orange snow in the winter of 2007. Locals feared the oily, smelly snow was the by-product of industrial pollution, a rocket launch, or maybe even a nuclear accident, but the blame eventually landed on a massive sandstorm in Kazakhstan.

In case it comes up: Wind chill = 35.74 + 0.6215T - 35.75(V^0.16) + 0.4275T(V^0.16). T is the air temp in °F; V is the wind speed in mph.

Washed Out Islands

A hurricane in New York is a pretty rare occurrence; they only hit about once every 75 years. In 1893, a Category 2 hurricane made landfall near present-day JFK Airport and caused extensive damage to the city, uprooting trees in Central Park, tossing wrought iron gates through buildings, and destroying nearly every building on Coney Island. The storm also obliterated a mile-long barrier island known as Hog Island, which was home to several saloons and bathhouses. It eroded the island and destroyed all of its buildings; a few years later it was reduced to a few mounds of sand. Oddly enough, this storm struck well before trendy hurricane names, so it was known only as the West Indian Monster of 1893. Researchers discovered dozens of antique items buried in the sand when the Rockaway Beach shores were being rebuilt in the 1990s.

Advanced Degrees

Midwesterners are accustomed to using both their heat and air-conditioning on the same day due to dramatic temperature changes and unseasonable weather. The "Great Blue Norther" of 1911 was the most dramatic cold snap ever recorded—several cities set record high and low temperatures on the same day. On November 11, 1911 (yes, 11/11/11) a massive storm system separated warm air from arctic air, yielding violent wind and storms. Kansas City, Missouri, reached a high temperature of 76°F, and by midnight, the temperature plunged to 11°F. The 65-degree difference was replicated in Oklahoma City, Oklahoma, and Springfield, Missouri.

In addition to the temperature changes, the front also caused dust storms, tornadoes, and blizzards from Oklahoma to Ohio. Nine people were killed by an F4 tornado in Janesville, Wisconsin; an hour later rescuers were working in near zero temperatures and blizzard conditions to rescue victims.

Raining Cats and Frogs

Yes, it has rained frogs in real life, not just in the movie *Magnolia*. Birds, bats, fish, and even worms have all reportedly fallen from the sky. Scientists theorize that fast-moving storms and waterspouts cross a body of water and sweep or suck up animals, then deposit them miles away. Residents of Honduras have celebrated the *Lluvia de Peces* (Rain of Fish) annually for more than a century. Some researchers have theorized that the hundreds of fish are sucked from the ocean and deposited 140 miles inland, while others have indicated that the fish may be pulled from underground water sources.

Animals have been known to survive the traumatic process, appearing startled but otherwise fine. But usually, they aren't so lucky, and don't survive the fall. Two instances in the 19th century indicate that cows were sucked up into the sky during a storm, and returned to earth in tiny pieces. Animals can also freeze to death in the frigid temperatures of the atmosphere, some of them are encased in ice when they make landfall.

The highest temperature ever recorded on Earth was 136°F — it happened in Al Aziziyah, Libya, in September 1922.

Disappearing Seasons

Volcanic winters, a phenomenon in which volcanic ash obscures the rays of the sun and increases the earth's reflectivity, cause dramatic decreases in temperature. In 1816, a volcanic winter led to a year where temperatures were so low in Europe and the United States that it was dubbed "The Year Without a Summer." Volcanic ash from several eruptions, including Mount Tambora in Indonesia, caused irregularities worldwide, but the effects were most severe in Europe, Canada, and the northern United States. A harsh frost in May destroyed many crops, snowstorms hit New England in June, and Pennsylvania's rivers and lakes iced over in July and August. Snow fell in tropical locales like Thailand, and colored freezing rain and snowfall blanketed Hungary and Italy.

This frosty weather had some lasting cultural influences. Mary Shelley and John Polidori went on a vacation to Switzerland with their friends but were forced to stay inside. To keep things interesting, they held a contest to develop the scariest story, leading to *Frankenstein* and *Vampyre*. Since the chilly weather decimated the feed sources for horses, German Karl Drais invented the velocipede, the predecessor of the modern bicycle.

To avoid sparking panics the U.S. Weather Bureau prohibited forecasts of tornadoes until 1938.

GET EXCITED!

FOUR APHRODISIACS FROM AROUND THE WORLD

1 Caffeine Cocktails

Since ancient times, most great sex has taken place when both parties were awake. Maybe that's why stimulants, from geisha tea to Red Bull, have long been held in high esteem as aphrodisiacs. According to a 1990 study in the *Archives of Internal Medicine*, drinking coffee increased sexual activity in 744 participating Michigan residents over the age of 60. The results strongly suggested that caffeine promotes arousal. That, or the subjects confused the study with a casting call for another sequel to *Cocoon*.

While caffeine has not yet been directly linked to an increased sex drive, the consensus in the medical community is that anything that gets the central nervous system pumping will have a general stimulating effect on the body. This explains why the ancient herb ginseng, which is said to increase energy and memory, is considered a strong aphrodisiac. It impacts the central nervous system, gonadic tissues, and the endocrine system, thus enhancing arousal. Ginseng has long been respected in China for its systemic healing properties, including the ability to aid sexual function.

Loco for Cocoa

The playground legend that green M&M's make you aroused has certainly made its rounds, but it's nothing more than an unsubstantiated myth. The green part of it, that is. Actually, chocolate is one of the most powerful edible aphrodisiacs in the world—and has been for quite some time. According to ancient Aztec history, 12 cacao beans (the beans used to make cocoa and chocolate) could purchase the services of a prostitute, and Montezuma reportedly downed 50 cups of liquid cacao to rev up before conjugal visits to his vast harem.

The scientific explanations for the arousing effects of chocolate are found in phenylethylamine (PEA) and anandamide (AEA). PEA is the chemical that causes elevated heart rates, increased energy, euphoria, and generally any symptom corresponding to feelings of being "in love." So, apparently, PEA is what makes us drive by our loved ones' houses late at night and compulsively scan our caller IDs. PEA's cohort, AEA, is a neurotransmitter that acts on the brain in a similar fashion to tetrahydrocannabinol (THC), the same chemical found in marijuana. And while chocolate won't get you stoned (sorry, dude), the presence of AEA probably explains chocolate's ability to calm and mellow.

In 2010, a sex pheromone found in male mouse urine was named "darcin," after the character Mr. Darcy in Jane Austen's *Pride and Prejudice*.

3 Laced with Paste

Sometimes, edible aphrodisiacs are never meant to be consumed, but rather smeared onto the body. In the ancient Arabian sex manual, *The Perfumed Garden*, rubbing the penis with various ointments is prescribed for "increasing the dimensions of members and making them splendid." Similar procedures are recommended in the *Kama Sutra*. Ingredients for such practices include honey, camel's milk, and lavender. While intriguing, the efficacy of the prescription probably has more to do with lubrication and the action of "repeatedly anointing the member" than the actual recipe.

An especially memorable recommended concoction for this instructs the man to catch a vulture by himself (very important) and mix the meat with honey and the juice of an amalaka (an Asian gooseberry-like fruit). Apparently, rubbing your body with dead vulture paste has the ability to bewitch the opposite sex, "even if a bath is taken afterward." How hot is that?

Red Hot Chili Peppers

Before Anthony and Flea, there were habaneros to get everyone hot and sweaty. For centuries, people have turned to chili peppers to spice up their love lives. In fact, in the 1970s, the Peruvian government, apparently fearing "big house love," banned chili sauce from prison food, declaring it inappropriate for "men forced to live a limited lifestyle."

The theory at work for this aphrodisiac is that chilis ignite in more ways than one. Think about what happens after you eat a big, mean chili pepper: Your palms sweat, your lips burn, and your breathing begins to shorten. One thing leads to another, and if your lover doesn't leave you for a big glass of milk . . . *arriba!* Another theory as to why searingly hot chilis arouse has to do with the pain they inflict. Pain causes the body to release endorphins, which try to block the signal of physical distress to the nervous system. These are the same kind of endorphins that are released during exercise and after sex, creating that feeling that all is right with the world. So masochists take note: If the whip is out of commission, then hit the Mexican produce stand.

Salvador Dalí wore a homemade scent made of fish glue and manure to help attract women.

MOONS OF THE NON-GLUTEAL VARIETY

FIVE FACTS FOR THE LUNAR ENTHUSIAST

1. No planet's moons can take a beating quite like Saturn's. The moon Mimas has a huge crater that is the result of a collision that almost split the moon in half; Hyperion has a squashed shape and rotates with a wobble for the same reason.

2. Venus has a "hypothetical moon"—an object observed by Giovanni Cassini in 1672 and backed up by 30 observations from 1740 to 1768. No one's seen any trace of the mysterious satellite since. Modern astronomers theorize that the disappearing moon may have been the result of optical illusions or stars that were in the vicinity of the planet.

3. Scientists briefly thought Mercury had a moon in 1974. Upon closer inspection, it turned out to be a star, 31 Crateris.

4. The names of Mars' moons Phobos and Deimos translate roughly to "fear" and "flight" in Greek. Though the names sound odd, there's actually a very reasonable explanation: Ares, the Greek equivalent of Mars, the Roman God of War, had sons named—you guessed it—Phobos and Deimos.

5. Every single one of Uranus's 27 moons are named for Shakespearean characters or characters from Alexander Pope's *The Rape of the Lock*. The names? We thought you'd never ask:

From *The Rape of the Lock*: Ariel, Belinda, and Umbriel

From Shakespearean works:

A Midsummer Night's Dream: Titania, Oberon, Puck
The Tempest: Miranda, Caliban, Sycorax, Prospero, Setebos, Stephano, Trinculo, Francisco, Ferdinand
King Lear: Cordelia
Hamlet: Ophelia
The Taming of the Shrew: Bianca
Troilus and Cressida: Cressida
Othello: Desdemona
Romeo and Juliet: Juliet, Mab
The Merchant of Venice: Portia
As You Like It: Rosalind
Much Ado About Nothing: Margaret
The Winter's Tale: Perdita
Timon of Athens: Cupid

Olivia Newton-John's grandfather was Max Born, one of the 1954 Nobel Prize winners for physics and the man who coined the term "quantum mechanics."

SCANDALOUS SCIENCE!

FIVE NOBEL PRIZE STORIES RIPPED FROM THE TABLOIDS

1 Scientist, Editor Duel Over Marie Curie's Honor! Gunshots Barely Avoided as Albert Einstein Weighs In

Just as her reputation as a brilliant scientist was growing, Marie Curie found herself at the center of a spectacular sex scandal.

Four years after her husband, Pierre Curie, died in a 1906 carriage accident, Marie became entrenched in a torrid love affair with one of his former students, physicist Paul Langevin. The two were sharing a love nest in Paris when Langevin's wife grew suspicious and decided to investigate. She hired a man to break into their pad and steal incriminating letters, which were then leaked to the press.

French newspapers went after the story with gusto. They painted Curie as a home-wrecker and a seductive Jew, even though she wasn't Jewish. The story played into the xenophobia of the time, and it fanned public outrage. The situation got so bad that one night, Curie returned home from a conference in Belgium to find an angry mob surrounding her house, tormenting her two daughters. She quickly packed up her family and fled to a friend's home.

Even now, Marie Curie's notebooks are too radioactive to be picked up by hand.

Eager to defend Curie's honor, Langevin challenged one of the newspaper's editors to a duel. The two men faced off against each other, but no one fired a shot. Meanwhile, another man came to Curie's defense. Albert Einstein offered a bit of reasoning that seemed both peculiar and offensive. He argued that Curie "has a sparkling intelligence, but despite her passionate nature, she is not attractive enough to represent a threat to anyone."

In 1911, at the height of the whole scandal, Curie won her second Nobel Prize. The Nobel committee suggested that she skip the awards ceremony, but she went anyway. The furor died down eventually, no doubt aided by Curie's humble demeanor and blinding dedication to science.

2 The Nobel Prize Sperm Bank—Deposits and Withdrawals

The Repository for Germinal Choice, better known as the Nobel Prize Sperm Bank, was founded in 1980 by multimillionaire Robert Graham, inventor of shatterproof eyeglass lenses. His goal was to combine the sperm and eggs of superior men and women—ideally Nobel laureates—to produce superior babies. If all this sounds an awful lot like eugenics, well, it was.

Most Nobel Prize winners were smart enough to steer clear of the bank, but three decided to make a deposit. One of these was William Shockley, who won the award for inventing the transistor and was an unapologetic racist. The other sperm donors were more random, and at least one of them lied about his intelligence.

But was the Repository for Germinal Choice a failure? That's hard to say. It brought more than 200 babies into this world, and many had higher-than-average IQs. In the end, however, its biggest legacy was that it changed how sperm banks work by offering detailed profiles of the donors. Now it's commonplace for women to choose the looks, professions, and interests of the men whose sperm they wish to use.

3 LSD-Using Nobel Laureate Speaks Out: Cites Belief in Ghosts, Disbelief in AIDS

Kary Mullis won his 1993 Nobel Prize in chemistry for inventing a way to quickly replicate DNA, which paved the way for modern genetic mapping and fingerprinting. But had the Nobel committee been giving out prizes for drug use and showboating, he probably would have won those, too.

A former LSD enthusiast, Mullis has no fear of sounding crazy. In his autobiography on the official Nobel website, for example, he includes a series of rambling anecdotes about the week he spent hanging out with the ghost of his dead grandfather. He's also quick to praise his own scientific achievements, even though he hasn't published a peer-reviewed paper since 1986. Arguably, he's the most controversial living Nobel laureate by sheer force of personality.

While his academic output has been limited, Mullis hasn't been shy about, er, enlightening the world with his theories. After receiving the Nobel, Mullis used his newfound status to talk publicly about topics far outside his field. Like, say, AIDS. Mullis doesn't believe

The FBI maintains a dedicated Art Crime Team of 13 special agents and three special trial attorneys.

in AIDS—at least not in the way most of the scientific community understands it. Instead, Mullis claims that AIDS is actually several diseases linked together by big pharmaceutical companies to make money. He also says that no one has ever proven the link between HIV and AIDS. (Mullis must have missed a paper in a 2003 *New England Journal of Medicine* that summarizes more than 25 years of independent peer-reviewed studies showing that AIDS is, in fact, a single disease caused by HIV.)

4 Doris Lessing Unfazed by Prize Patrol: Celebrated Author Claims, "I couldn't care less."

In 2007, British writer Doris Lessing, best known for her 1962 feminist masterpiece *The Golden Notebook*, won the Nobel Prize in literature. Thing is, she wasn't exactly gracious about her win. She told the hordes of reporters camped outside her door, "I couldn't care less." Lessing then elaborated on why she was so unmoved: "I'm 88 years old, and they can't give the Nobel to someone who's dead, so I think they were probably thinking they'd probably better give it to me now before I've popped off."

5 Economist Loses $4 Billion, Sparks Worldwide Financial Meltdown

Myron Scholes, cowinner of the 1997 Nobel Prize in economics, has been called "the intellectual father of the credit-default swap." Scholes won the award for what's known as the Black-Scholes method for determining the value of derivatives and stock options. Over the years, his framework became the standard for financial markets and was even referred to as the Holy Grail of Economics.

Recently, however, all of that has changed. Although there was a lot of hype about the theory, the Black-Scholes model was dangerous when used incorrectly. In fact, it's largely what defined the credit-default swaps that were behind the mortgage-backed securities that helped trigger the recent global economic crisis. Even as Scholes was being awarded the Nobel Prize, things weren't going too well for him. The hedge fund he'd founded was about to lose an astronomical $4 billion in four months.

But Scholes still believes in his theory. In an interview with the *New York Times* in May 2009, journalist Deborah Solomon asked, "In retrospect, is it fair to say that the idea that banks could manage risk was a total illusion?" Scholes responded, "What you're saying is negative. Life is positive, too. Every side of a coin has another side."

What exactly that other side is, Scholes didn't mention.

A week before the stock market crash of 1929, a Yale economist said, "Stocks have reached what looks like a permanently high plateau."

THE SCIENCE OF ART

FIVE WAYS MUSEUMS DETECT FORGERIES

Think artists and scientists don't have anything in common? Think again. Here are just five hi-tech ways the scientists and curators team up to debunk forged artworks.

1 Cracking the Craquelure

After years of straining against its frame, the canvas of an old painting slowly starts to lose its elasticity. The paint on it, though, isn't quite as flexible; as the canvas stretches, the painting itself starts to crack. These tiny breaks, known to art restorers as craquelure, aren't random, though. Older paintings tend to have more, and there are generally more cracks closer to the frame where the canvas has stretched more. Different paints and canvas-stretching techniques cause different craquelure patterns, thus there are said to be distinct French, Italian, and Dutch types of cracking.

Pablo Picasso was held, questioned, and released in 1911 for his suspected role in the theft of the Mona Lisa, which turned up months later.

Forgers know this, and go to great lengths heating and drying their paintings in an effort to match the craquelure patterns of the era and region they're copying. The forger behind "The Virgin and Child with an Angel" was not willing to take any chances. Supposedly painted by Francesco Francia in 1490, this piece from the National Gallery came under scrutiny when an almost identical painting showed up at an auction. When museum staff put the canvas under a microscope, they realized that what they thought was craquelure had actually been painted onto the canvas by the forger.

2 X-raying the Fake Virgin

When London's Courtauld Gallery was bequeathed the art collection of Lord Lee of Fareham, their curators could hardly believe their luck. Tucked in among various Italian paintings was what they thought was an original 15th century piece from the Florentine genius Botticelli. "Madonna of the Veil," hailed by critics as a masterpiece, was the star of the gallery, and would have remained so were it not for Kenneth Clark, the director of the National Gallery at the time.

Clark noticed that the Virgin of that painting was, well, sexier than her other representations in the Botticelli oeuvre. Her full lips, in fact, reminded Clark of a very different leading lady—the movie star Jean Harlow. The painting wasn't fully discredited, though, until 1994, when an energy dispersive X-ray microanalysis, which determines the elements in a paint sample, found that the paint colors included distinctly modern colors like cobalt blue and opaque chromium oxide green, the latter of which wasn't available until 1862. The former Botticelli masterpiece is now thought to have been created by the art instructor Umberto Giunti sometime in the late 1920s—right around when Harlow made her silver screen debut.

3 Looking Inside Painting Layers

Polish scientists have repurposed a tool for studying eye disease to detect art forgery. Optical coherence tomography (OCT) is basically a high-res ultrasound that uses light waves instead of acoustic waves to create cross-sectional images of various objects. Because it's a non-invasive way to look inside fragile objects, doctors have used it to track macular degeneration in people's retinas.

But there are other things besides biological tissues that have delicate layers. Old paintings, for example, often consist of layers of glue, paint, glazes, and varnish. The Polish physicists soon found that OCT could see inside of them as well. Before OCT tomographs, curators used to study a suspect area of a painting by cutting off a small section from it and looking at it under a microscope to see in what order the layers were added.

One of the first paintings they studied, the 19th-century painting "Portrait of an Unknown Woman," showed off the tool's promise brilliantly. The OCT tomography quickly revealed what other forms of examination could not—that someone tampered with the artist's signature. While the forger was skilled enough to revarnish the entire painting

Between 1912 and 1948, art competitions were a part of the Olympics. Medals were awarded for architecture, music, painting, and sculpture.

after adding the new signature, the tomography showed the deception. Not only was the signature painted in an area where the primary varnish had been removed; it was actually on top of some of the remnants of that varnish, showing that it was an intentional forgery.

4 Guilt by Manganese

Here's a lesson for forgers: Stick to your guns. In the early 1900s, Riccardo Riccardi and his friend Alfredo Fioravanti realized that there was some serious money to be made in selling antique Etruscan pottery to museums. Since they didn't actually have any antiques lying around, they decided to make some themselves. They started by building giant warrior statues. Because they didn't have kilns large enough to fire them whole, they broke them and cooked them in pieces, leaving a few pieces out to make it look like they really were antiques.

The warriors sold for exorbitant prices to major museums, but it wasn't until the Metropolitan Museum of Art displayed them decades later that the first doubts arose about their authenticity. Experts saw the warriors as suspicious because the pieces lacked vents for firing and because they didn't quite match up stylistically with other statues from the time.

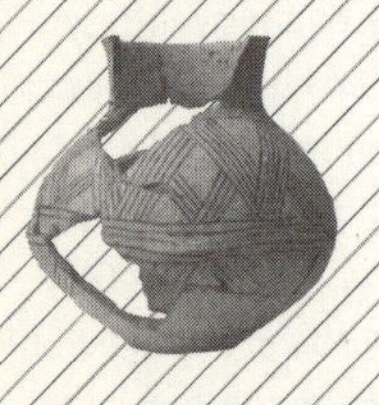

Amid growing doubts the museum hired John Veach Noble to do some tests using a technique called X-ray fluorescence, which bombards material with gamma rays until it emits radiation characteristic of the elements inside. He found manganese in the clay, which wasn't present in other Etruscan sculptures of the era. Faced with the mounting evidence, Fioravanti admitted guilt, and even produced one of the warriors' thumbs that he had kept as evidence. Had Fioravanti kept quiet, though, he may have stayed clean; soon after he admitted his fakery, scientists found manganese in real ancient Etruscan statues.

5 Thermoluminescence Captures the Essence

When Oxford Authentication, an antique certification lab in London, first received a purportedly ancient Chinese horse statue for testing, the results were encouraging. The first piece of clay they tested using thermoluminescence dating glowed incredibly brightly as it heated, showing that it had absorbed enough environmental radiation over the years to make it more than a thousand years old. Thus, the horse seemed to be an incredibly valuable sculpture from China's Tang Dynasty.

Unfortunately, as the technicians tested clay from other areas of the sculpture, they soon realized the truth—the equine masterpiece was a modern fake. While the deception wasn't unusual—more than 80 percent of ancient Chinese artifacts on the market are thought to be fraudulent—the trickster's technique was. Knowing that testers often take clay samples from unseen areas of sculptures to protect the pieces' aesthetics, the forger had built the new sculpture on an ancient base, banking that the technicians would test the latter and not the former.

The clean, pleasant smell that accompanies rain falling on dry ground has a name—it's called petrichor.

10 LISTS

YOU CAN SHARE WITH YOUR KIDS (OR YOUR INNER CHILD)

FIVE FAIL-SAFE RITUALS FOR PROTECTING YOUR NEWBORN

Jump Over Your Baby

In parts of northern Spain, newborns participate in a ceremony that seems part Olympic track-and-field event and part Evel Knievel stunt. Several babies are placed on a mattress while a man long-jumps over them. The ceremony is based on the biblical story in which King Herod ordered all male babies in the area to be killed after hearing that a "new king" has been born in Bethlehem. Just as Mary and Joseph escaped with baby Jesus to Egypt, this Spanish ritual is meant to symbolize a similarly dangerous experience for a child. After braving the danger and emerging (hopefully) unharmed, the child is prepared for a safe passage through life.

No U.S. president was an only child.

2 Play Some Baby Hot Potato

In Bali, many natives observe a custom whereby the baby isn't allowed to touch the ground (or cradle, or whatever) for the first 105 days of the child's life. Instead, the newborn is continuously held by family members.

3 Smoke Your Baby

In Kimberley, Australia, many Aboriginal mothers still practice the art of "baby smoking." The ritual is meant to protect the child by giving him the blessing of the tribal mothers in addition to the baby's "earth mother." Branches and leaves from sacred konkerberry shrubs are burned, creating what are believed to be purifying fumes. Then the mother squeezes her breast milk into the fire, and the grandmother waves the baby through the smoke.

Don't Name Your Baby

Many societies believe that newborns are particularly susceptible to evil spirits, and a baby's name is sometimes kept secret (or not given at all) so it can't be used against the child in spells. In some Haitian, Nigerian, and Romani cultures, babies are given two names at the time of birth. Parents keep one name a secret and don't share it with the child until he's considered old enough to guard the name for himself.

Similarly, in Thailand a newborn is often referred to by a nickname to escape the attention of evil spirits, who are believed to be the ghosts of dead, childless, unmarried women. The newborn is given a two-syllable name, which is mainly used later on by teachers, employers, and during formal occasions. Some Vietnamese parents even delay naming their baby until he's more than one month old—the safety margin, spirit-wise. They also discourage anyone from complimenting the newborn; instead, they refer to the tot as "ugly" or "rat" to deter evil spirits, who prefer harassing attractive babies.

5 Cut the Cord (Then Bury It in a Special Place)

Overprotective moms, take note! The Navajo tribe of Native Americans believe that if the umbilical cord and placenta of a newborn are buried near the family's house, the child will always return home. The placenta is also sometimes buried next to objects that symbolize the profession the parents hope their child will pursue, which may explain the spike in buried stethoscopes found all across the land.

The wife of Atlanta Hawks GM Pat Williams had a baby on the NBA's draft day in 1974. To celebrate, Williams drafted his newborn son in the 10th round.

RAW UMBER GOT A RAW DEAL

FIVE TIMES CRAYOLA FIRED ITS CRAYONS

Cousins Edwin Binney and C. Harold Smith introduced their first eight Crayola crayons in 1903. Since then, the world has changed, and so, too, have the names of their waxy creations. Be it ever-shifting societal, racial, or political atmospheres, these crayons of yore have a revisionist history unto themselves.

"Flesh" Crayons Change Their Name

While everyone acknowledges that the civil rights movement brought about great strides in American society, most individuals overlook the huge advances it brought to the crayon community. In 1962, Crayola voluntarily changed "Flesh" to "Peach" in an attempt to avoid any legal issues and encourage people to embrace seeing the world in black and peach.

Crayola means "oily chalk." The name combines "craie" (French for "chalk") + "ola" (short for "oleaginous," or "oily").

2 Prussian Blue Receives Icy Treatment

The Kingdom of Prussia (part of modern-day Germany and Poland) remained an independent state from 1701 to 1871, but the crayon dubbed Prussian Blue had a far shorter reign in the kingdom of colors. Introduced in 1949 alongside a cadre of 39 new cohorts, Prussian Blue was unceremoniously stripped of its name in 1958, after teachers continued to voice concerns that the crayon wasn't Cold War–sensitive. (Plus, kids didn't know where Prussia was anymore!) Crayola hoped the color's new name, Midnight Blue, would help make it less political and certainly less useful in coloring Iron Curtains.

3 Indian Red Was a Nod to India?!

Introduced in 1958 with 15 additional colors (finally giving children 64 shades to work with!), this color was actually named for a pigment that originated in India. Over the years, teachers began to worry that children would see the crayon as a reference to American Indians' skin color. In 1999, the Crayola company changed the name to Chestnut—but that too came with a disclaimer. The crayon manufacturer warned children that, despite the famous song, these chestnuts should never be roasted over an open fire. Mainly because they soften and melt at around 105°F.

4 Eight Men Out: Colors Get Waxed Off

The year 1990 brought about the first forced retirement of colors in the house of Crayola. And just like that, old fogies Blue Gray, Green Blue, Lemon Yellow, Maize, Orange Red, Orange Yellow, Raw Umber, and Violet Blue were sent out to waxy pastures. They were replaced with new-generation colors including Cerulean, Fuchsia, and Dandelion, which were considered bolder, more vibrant, and more likely to boost your Scrabble score.

5 Kindergarteners Get Drunk with Power

In celebration of Crayola's 100th birthday in 2003, consumers were encouraged to suggest new crayon names as well as vote out four crayon colors. The casualties of the Crayola tribal council were newer colors Blizzard Blue, Magic Mint, and Teal Blue, and the older Mulberry. These proud veterans stepped aside for such wildly creative crayons as Inch Worm, Jazzberry Jam, Mango Tango, and Wild Blue Yonder—proving that allowing kindergarteners to have veto power over your marketing department isn't always the best idea.

According to Crayola, the average kid wears out about 730 crayons by his or her 10th birthday.

SLINKING INTO OUR HEARTS

HOW SEVEN CLASSIC TOYS WERE INVENTED

1. Lincoln Logs are the brainchild of John Lloyd Wright, Frank Lloyd Wright's son. The original instructions included a how-to on constructing a replica of Abraham Lincoln's cabin, as well as directions on how to build Uncle Tom's cabin.

2. A stonemason invented Tinkertoys after he saw kids building things with pencils and spools of thread, and being thoroughly entertained by it.

3. Hula Hoops have been around forever in various formats, but the "official" Wham-O toy debuted in 1958. The inventors promoted it by going around to various playgrounds and parks giving children samples and showing them how to use it. Something tells us two random men showing up in a park handing out toys wouldn't go over that well today . . .

4. Play-Doh was first sold as a wallpaper cleaner. Homeowners rolled it on the walls to remove coal dust.

5. Troll dolls were created in 1949 by a Danish fisherman who needed a cheap Christmas gift for his daughter because he couldn't afford to buy anything. He used sheep's wool for the hair. Thomas Dam's dolls caught on; thus the original dolls were called Dam Dolls.

6. Slinky was the handiwork of Naval engineer Richard James. He was working to develop springs that could keep ship instruments stable in choppy waters when he knocked one of the springs off its shelf. The spring did what a Slinky does . . . it stepped down to a stack of books, then to the table, and then to the floor, where it righted itself into a cylinder. James knew it would be a great toy and tests by neighborhood kids proved him right.

7. The sock monkeys that we have come to know and love today—the ones made with Red-Heel socks—are thought to have come about in 1932. The distinctive red heel (the monkey's mouth) was given to the socks so customers would know they were getting authentic Rockford socks. When the Nelson Knitting Company discovered that their socks were being used across the country in this arts-and-crafts movement, they won the design patent for the sock monkey pattern and started including it in the packaging of their socks.

In 2008, the Little Tikes Company sold 457,000 Cozy Coupes, making it the bestselling car in America.

TOY SOLDIERS

FOUR KIDS' TOYS THAT HAVE GONE TO WAR FOR AMERICA

1 Slinky

No toy seems to exhibit a fear of heights the way the Slinky does. Place the cowardly coil on the top of a staircase, and it immediately starts inching its way down. But in battle, this spiral wonder has proven far braver. In Vietnam, the toys worked from treetops. Radio operators would tie rope through the middle of the long metal spirals, then drape them over branches to create perfect radio antennas. The Slinkys were especially useful because they didn't tangle and could be yanked down quickly when soldiers needed to run. Of course, if you're off to battle, buy an extra. While one Slinky will do the job, most radio operators wielded two for better reception.

In 1999, the NSA's Maryland HQ banned Furbies from its premises because it was feared they could repeat national security secrets.

Metal Crickets

During the D-day invasion at Normandy, more than 15,000 men dropped from the night sky carrying little metal insects. At the time, flashlight signaling was a common way for soldiers to communicate, but officials worried that flashing lights could tip off observant Germans. So, they substituted the torches for cheap, wind-up crickets that made clicking sounds. Each paratrooper received one, along with instructions to identify himself through the chirps. The brilliant scheme helped soldiers meet up safely on the ground.

Unfortunately, the success was short-lived. A few Germans caught on, and after capturing some of the crickets, they used them to trap unsuspecting Americans. Still, the toys contributed to the invasion's success, and souvenir replicas are still sold to tourists in Normandy today.

Silly String

Since the start of the Iraq war, worried families have sent tens of thousands of Silly String cans to their sons and daughters in the military. Why? Shortages of night-vision goggles have forced soldiers to improvise with the fluorescent foam. Silly String is especially good for finding trip wires in the dark. Before entering suspicious rooms, soldiers spray the string everywhere. If the glowing blobs hang ominously in midair or get stuck on previously invisible objects, soldiers know to tread lightly.

The View-Master

The View-Master was originally for grown-ups. Developed in the late 1930s, the device was used as a training tool to help WWII soldiers recognize specific ships, planes, and artillery from afar. But after the war, the device got a kid-friendly makeover. The toy was outfitted with reels featuring television characters and tourist destinations, and it quickly became a staple in toy chests everywhere.

In 1990, engineers trying to fix an antenna problem on the Hubble Space Telescope used Tinkertoys to build a partial model.

FOUR COMIC SUPERHEROES

WHO MADE A REAL-WORLD DIFFERENCE

Popeye Helps America Survive the Great Depression

Everyone knows Popeye's secret. Whenever the cartoon sailor is on the verge of losing a fight, he squeezes open a can of spinach, pours the greens down his throat, and uses his supercharged muscles to pummel opponents. But fewer people know that the U.S. government is directly responsible for his dependence on canned vegetables.

In the 1930s, America was mired in the Great Depression, and the government was looking for a way to promote iron-rich spinach as a meat substitute. To help spread the word, they hired one of America's favorite celebrities, Popeye the Sailor Man. It was a smart plan. In all of the comic strips up to that point, Popeye's superhuman strength had never been explained. But with the government's campaign in place, Popeye was suddenly more than willing to share the secret to his strength. Sure enough, soon after Popeye took up spinach, American sales of the mighty veggie increased by one-third. Better still, American children rated it their third favorite food, right after turkey and ice cream.

But it wasn't just spinach the government was endorsing. They were also pushing the merits of canned food. U.S. officials wanted Americans to know that cans were the perfect way to stock up on emergency rations.

2 Superman Defeats the Ku Klux Klan

In the 1940s, "The Adventures of Superman" was a radio sensation. Kids across the country huddled around their sets as the Man of Steel leaped off the page and over the airwaves. Although Superman had been fighting crime in print since 1938, the weekly audio episodes fleshed out his storyline even further. It was on the radio that Superman first faced kryptonite, met *The Daily Planet* reporter Jimmy Olsen, and became associated with "truth, justice, and the American way." So, it's no wonder that when a young writer and activist named Stetson Kennedy decided to expose the secrets of the Ku Klux Klan, he looked to a certain superhero for inspiration.

In the post–World War II era, the Klan experienced a huge resurgence. Its membership was skyrocketing, and its political influence was increasing, so Kennedy went undercover to infiltrate the group. By regularly attending meetings, he became privy to the organization's secrets. But when he took the information to local authorities, they had little interest in using it. The Klan had become so powerful and intimidating that police were hesitant to build a case against them.

Slumber Party Barbie of 1965 came with her own "How to Lose Weight" book. One of the tips: "Don't eat."

Struggling to make use of his findings, Kennedy approached the writers of the Superman radio serial. It was perfect timing. With the war over and the Nazis no longer a threat, the producers were looking for a new villain for Superman to fight. The KKK was a great fit for the role. In a 16-episode series titled "Clan of the Fiery Cross," the writers pitted the Man of Steel against the men in white hoods. As the storyline progressed, the shows exposed many of the KKK's most guarded secrets. By revealing everything from code words to rituals, the program completely stripped the Klan of its mystique. Within two weeks of the broadcast, KKK recruitment was down to zero. And by 1948, people were showing up to Klan rallies just to mock them.

The world's single largest piece of gum was made for Willie Mays to split up and give to sick kids. It equaled 10,000 pieces of Bazooka.

3 Captain Marvel Jr. Saves the Bad-Hair Day

Like most American kids in the 1940s, Elvis Presley fantasized about growing up to be like his favorite comic book superheroes. But it turns out that The King might have been more interested in their fashion statements than their special powers.

During his early teen years, Elvis was obsessed with Captain Marvel Jr., known as "America's most famous boy hero." A younger version of Captain Marvel, the character sported an unusual hairstyle that featured a curly tuft of hair falling over the side of his forehead.

Sound familiar? When Elvis set out to conquer America with his rock 'n' roll ways, he copied the 'do, thus making it one of the most famous hairstyles of the 20th century. But that wasn't all. Captain Marvel also gets credit for the short capes Elvis wore on the back of his jumpsuits, as well as The King's famous TCB logo, which bears a striking resemblance to Marvel's lightning bolt insignia.

Of course, Elvis never tried to hide his love for the Captain. A copy of Captain Marvel Jr. #51 still sits in his preserved childhood bedroom in an apartment in Memphis, and his full comics collection remains intact in the attic at Graceland. Plus, the admiration was mutual. Captain Marvel Jr. paid tribute to The King in one issue, referring to the singer as "the greatest modern-day philosopher."

4 A Spider-Man Villain Keeps Folks Out of Jail

In a 1977 edition of *Spider-Man*, Peter Parker has the tables turned on him. The villain, Kingpin, tracks down Spidey using an electronic transmitter that he'd fastened to the superhero's wrist. Although Kingpin loses in the end (he always does), one New Mexico judge saw beauty in his plan. Inspired by the strip, Judge Jack Love turned to computer salesman Michael Goss and asked if he could create a similar device to keep track of crime suspects awaiting trial. In 1983, Goss produced his first batch of electronic monitors. Authorities in Albuquerque then tested the devices on five offenders, using the gadgets as an alternative to incarceration. Today, the transmitters are a common sight in courtrooms across the country, usually in the form of electronic ankle bracelets. Most famously, Martha Stewart donned one while she was under house arrest in 2004. Perhaps she would have felt better knowing that the gadget had once nabbed Spider-Man, too.

Smokey the Bear's original name was Hot Foot Teddy.

ON THE ONE HAND

THE STORIES BEHIND NINE MUPPET FAVORITES

1 Kermit

Kermit was "born" in 1955 and first showed up on "Sam and Friends," a five-minute puppet show by Jim Henson. The first Kermit had humble origins: Henson used one of his mom's coats and some Ping-Pong balls to build him. At the time, Kermit was more reptilian than frog-like. By the time he showed up on *Sesame Street* in 1969, though, he had made the transition to frog. There are rumors that he got the name Kermit from a childhood friend of Henson's or a puppeteer from the early days of the Muppets, but Henson always refuted both of those stories.

2 Oscar the Grouch

Caroll Spinney voices both Big Bird and everyone's favorite green curmudgeon. Spinney has said he based Oscar's cranky voice on a particular NYC cabdriver he once had the pleasure of riding with. Oscar was originally an alarming shade of orange before transitioning to his familiar green. But the trash can is just his American pad. In Pakistan, his name is Akhtar and he lives in an oil barrel. In Turkey, he is Kirpik and lives in a basket. And in Israel, it's not Oscar at all—it's his cousin, Moishe Oofnik, who lives in an old car.

Cookie Monster

Jim Henson drew some monsters eating various snacks for a General Foods commercial in 1966. The commercial never aired, but Henson recycled one of the monsters (the "Wheel-Stealer") for an IBM training video in 1967 and again for a Fritos commercial in 1969. By this time, he started working on *Sesame Street* and decided the voracious monster would have a home there.

4 Elmo

According to *Sesame Street* writers, an extra red puppet was just lying around. People would try to do something with him, but nothing really panned out. In 1983, puppeteer Kevin Clash picked up the red puppet and started doing the voice and the personality and it clicked—thus, Elmo was born.

In 1970, Ernie of *Sesame Street* reached #16 on the Billboard Hot 100 with the timeless hit "Rubber Duckie."

5 Count von Count

The Count made his first appearance in 1972 and was made out of an Anything Muppet pattern—a blank Muppet head that could have features added to it to make various characters. He used to be more sinister; he was able to hypnotize and stun people and he laughed in a typical scary-villain fashion. The writers quickly revamped the character to make him more appealing to kids.

6 Rowlf the Dog

Surprise, surprise, Rowlf debuted in 1962 for a series of Purina Dog Chow commercials. He went on to claim fame as Jimmy Dean's sidekick on *The Jimmy Dean Show* and was a series regular from 1963 to 1966. Dean said Rowlf got about 2,000 letters from fans every week. Rowlf nearly moved to *Sesame Street* but ended up becoming a regular on *The Muppet Show* in 1976.

7 The Swedish Chef

Real Swedish chef Lars "Kuprik" Bäckman claims he was the inspiration for the Swedish Chef. Bäckman says he caught Jim Henson's eye during a segment on *Good Morning America*. Henson supposedly bought the rights to the *Good Morning America* recording and created the Swedish Chef (who DOES have a real name, but it's unintelligible). One of the Muppet writers, Jerry Juhl, has said that in his years with Jim Henson on the Swedish Chef, he never heard that the character was based on a real person.

8 Telly Monster

Telly was originally the Television Monster when he debuted in 1979. He was obsessed with TV and his eyes would whirl around as if hypnotized whenever he was in front of a set. After a while, producers started worrying about his influence on youngsters, so they changed him to make him the chronic worrier he is now.

Miss Piggy

Miss Piggy's first TV appearance was actually on a Herb Alpert special. It wasn't until 1976, when *The Muppet Show* premiered, that she became the glamorous frog-loving blonde that we know and love today. Puppeteer Frank Oz once said that Miss Piggy grew up in a small town. Since her dad died when she was young and her mother was mean, Miss Piggy had to enter beauty contests to make money.

Ashrita Furman's 12-minute mile doesn't sound all that impressive—until you realize he was riding a pogo stick at the time.

BURPY, HICKEY, CHESTY, AND SEVEN OTHER REJECTED DWARF NAMES

Think naming seven dwarves is easy? Think again. Walt Disney and company brainstormed dozens of names before settling on Snow White's final entourage. A few of the rejects:

1. Jumpy
2. Deafy
3. Chesty
4. Hickey
5. Wheezy
6. Baldy
7. Gabby
8. Awful
9. Tubby
10. Burpy

Mr. Potato Head is the official travel ambassador for Rhode Island, where Hasbro is headquartered.

STEP ASIDE, NEWTON

SIX LAWS OF CARTOON PHYSICS

You may have aced high-school physics, but that won't help you understand cartoons. The animated universe has its own set of physical rules, honed over the years to maximize hilarity while maintaining a wacky sense of honor and order. Here are a few of our favorite scientific laws from that world.

1 Violent Death Is Impermanent

No matter how many times or ways in which Kenny McCormick gets killed in *South Park*, he's always back at the bus stop and back to life sooner or later. Same goes for both cat and mouse in *Tom and Jerry* and *The Simpsons'* meta-cartoon *The Itchy & Scratchy Show*. Usually, this miraculous reincarnation is accompanied by total amnesia from the resurrected one and the rest of the cast, and life returns to normal—or at least the way it was previously.

That cartoons fail to respect the permanence of death is probably related to the fact that they have extremely adaptable bodies. The *Family Guy*'s Peter Griffin, for example, has survived both extreme nickel poisoning and losing all of the bones in his body and emerged no worse for the wear. After all, as Roger Rabbit learned, the only real way to kill a toon is to dissolve him in Dip—a combination of turpentine, acetone, and benzene.

2 Weapons Don't Work Well

As Elmer Fudd should know by now, a cartoon gun is basically worthless. Even when it's not breaking, sagging, or popping out signs that say "bang," the fictional firearm is almost never lethal even when it fires correctly. Perhaps the rugged physiology and lightning fast reflexes of cartoon characters make them harder to kill. More likely, though, the guns just aren't that powerful. Even when fired from close range, they usually just blacken the victim's face, or, in the most extreme cases, rearrange his bill.

3 Bags or Coats Contain Multitudes

If you're animated, it's best to never leave the house without a jacket or satchel. That's because these objects all seem to possess four-dimensional polygons, known to physicists as tesseracts, that appear tiny but are actually large enough to house objects bigger than the wearer himself. How else could you explain that Dora the Explorer's backpack has produced both ladders and spacesuits or that in *Despicable Me*, Gru's pea coat can produce WMDs? Also, we're not sure if Bender's internal compartment on Futurama counts as either a bag or

In 2010, *Forbes* estimated Scrooge McDuck's net worth to be $33.5 billion. Good year for gold.

a coat, but it still seems to be able to hold everything from popcorn poppers to beer kegs to a baby grand piano.

4 Space Is Just a Construct

It's not just the characters' accessories that are wacky. At certain times, the toons themselves have access to hammerspace; that is, they can quickly enter and exit alternate dimensions. The term "hammerspace" comes from Japanese manga, where female characters often produce—out of thin air—oversized rice mallets to bonk the heads of men who offend them. Hammerspace only occurs at certain times or in certain places; most notably, there seem to be pockets of it behind thin trees and lampposts, allowing large men to hide in tiny spaces.

5 Gravity Is Relative

One of the most common tropes of the *Looney Tunes* series is this: A character walks or runs over a cliff but does not fall. Only when he finally notices his predicament and looks down does gravity finally take effect, and he plummets to the basin below at $9.8 m/s^2$. Given that this momentary suspension of physics most often happens during a pursuit of the Road Runner, science writer Jacqueline Houtman coined the neologism *Coyotus Interruptus* to describe it. The fungibility of gravity in the animated universe also manifests itself in another way. When cartoon characters are scared—or poked in the bottom—they often shoot straight upward as if rocketing away from a zero-G planet.

6 Cartoons Leave Cartoon-Shaped Holes in Industrial Materials

Right before the famous couch gag on the opening sequence of *The Simpsons*, Homer literally goes through a door in the family's garage, leaving what's known in the cartoon physics field as a "silhouette of passage"—a perfect outline of himself in the solid material he went through. While in reality it would be very difficult to find a material that works with humans—maybe a wall of Nutty Bars?—in the Tooniverse pretty much any substance can be perforated in this manner, provided that the penetrating agent is sufficiently scared or excited.

Fall and aftermath aside, the nursery rhyme never states that Humpty Dumpty is an egg.

SEVEN REASONS

MISTER ROGERS WAS THE BEST NEIGHBOR EVER

1 Even Koko the Gorilla Loved Him

Most people have heard of Koko, the Stanford-educated gorilla who could speak around 1,000 words in American Sign Language and understand about 2,000 in English. What most people don't know, however, is that Koko was an avid *Mister Rogers' Neighborhood* fan. As *Esquire* reported, when Fred Rogers took a trip out to meet Koko for his show, not only did she immediately wrap her arms around him and embrace him, she did what she'd always seen him do onscreen: She proceeded to take his shoes off!

Looking for Mister Rogers' actual neighborhood? The show filmed in Pittsburgh.

He Made Thieves Think Twice

According to a *TV Guide* profile, Fred Rogers drove a plain old Impala for years. One day, however, thieves swiped his ride from the street near the TV station. When Rogers filed a police report, every newspaper, radio, and media outlet around town picked up the story. Amazingly, within 48 hours the car turned up in the exact place from which it had been stolen, with an apology on the dashboard. The note read, "If we'd known it was yours, we never would have taken it."

He Saved Both Public Television and the VCR

Strange but true. When the government wanted to cut Public Television funds in 1969, the relatively unknown Mister Rogers went to Washington. Almost straight out of a Frank Capra film, his five- to six-minute testimony on how TV had the potential to give kids hope and create more productive citizens was so simple but passionate that even the most gruff politicians were charmed. While the budget should have been cut, the funding instead jumped from $9 to $22 million. Rogers also spoke to Congress and swayed senators into allowing VCRs to record television shows from the home. It was a contentious debate at the time, but Rogers' argument was that recording a program like his allowed working parents to sit down with their children and watch shows as a family.

❹ He Might Have Been the Most Tolerant American Ever

Mister Rogers seems to have been almost exactly the same offscreen as he was onscreen. As an ordained Presbyterian minister, and a man of tremendous faith, Mister Rogers preached tolerance first. Whenever he was asked to castigate non-Christians or gays for their differing beliefs, he would instead face them and say, with sincerity, "God loves you just the way you are." Often this tolerance provoked ire from fundamentalists.

He Could Make a Subway Car Full of Strangers Sing

Rogers was once running late to a New York meeting and couldn't find a cab, so he and one of his colleagues hopped on the subway. *Esquire* reported that the car was filled with people, and Rogers assumed they wouldn't be noticed. But when the crowd spotted Rogers, they all simultaneously burst into song, chanting "It's a beautiful day in the neighborhood." The result made Rogers smile wide.

He Composed All the Songs on the Show, and over 200 tunes.

❼ **Those Sweaters Had a Pedigree.** Many of the cardigans Rogers wore on the show had been hand-knit by his mother.

Cardigan sweaters are named for James Brudenell, 7th Earl of Cardigan, who wore the garment during the Crimean War.

LAWS OF THE MAGIC KINGDOM

SEVEN THINGS DISNEY PARKS HAVE BANNED

Disneyland may be the Happiest Place on Earth, but don't think that means you can just waltz in and do whatever you want. Here are just a few of the things on which the Mouse has dropped his hammer.

Long Hair

Until the late 1960s, men could either have flowing locks or enjoy Adventureland, but they definitely couldn't do both. According to Snopes.com, if a long-haired fellow tried to buy a ticket, a cast member would discreetly and politely inform the man that his hairdo didn't jibe with the park's unwritten dress code before escorting him from the premises.

Facial Hair

It's tough to find a picture of Walt Disney without a mustache, but for decades it was even tougher to find a Disney employee who had a 'stache of his own. Starting in 1957, workers at Disney parks were not allowed to have long hair, grow beards, or wear mustaches. (The underlying logic was that park patrons wouldn't want to buy a $9 soda from some filthy bearded hippie or mustachioed Snidely Whiplash type.)

In 2000, Disney was having trouble drumming up enough manpower to staff its parks, so it relaxed the facial hair ban. Employees were finally allowed to grow mustaches, provided they kept them trimmed and groomed. Beards didn't fare so well, though; they stayed on the forbidden list.

Blake Lively

How could anyone not like the cute-as-a-button star of *Gossip Girl*? Disneyland apparently wasn't always amused with Lively's pre-fame antics. According to Lively, when she was six, she and her older brother used the old put-hairspray-on-a-friend's-readmission-hand-stamp-to-transfer-the-stamp-to-their-own-skin trick. It would have been the perfect crime, except security nabbed the Lively kids right as they went through the park's turnstiles and slapped the pair with a one-year ban.

In 1991, Wayne Allwine, the voice of Mickey Mouse, married Russi Taylor—the voice of Minnie.

4 Costumes

You may want to dress up like Jack Sparrow for a day of riding Pirates of the Caribbean, but if you're older than 9, forget it. Disney bans any costumes and masks on anyone who's 10 or older. Also listed on Disney's park dress code: "Makeup that could be construed as part of a costume." So go easy on the eye shadow—the fashion police might decide you're shooting for a 19th-century harlot look and give you the heave-ho.

Similarly, the dress code bans "clothing that accentuates or draws attention to private areas," a rule whose odd phrasing undercuts its good intentions. Here's hoping Disney starts handing out unisex burlap smocks at the park gates to avoid any potentially accentuated private areas.

5 Kids

Kids banned by Disney? You bet. In January 2008, Disney announced that children under the age of 10 would no longer be allowed to dine at Victoria & Albert's, the ritziest restaurant at Disney World's Grand Floridian Resort and Spa. The move made news, but Disney officials claimed that the AAA five-diamond-rated restaurant didn't attract that many children in the first place. In addition to being pricey, Victoria & Albert's only offered a fixed-price menu with kid-unfriendly offerings like caviar. As a result, only a handful of young diners ate there each year.

6 Gum

Want to chomp on some gum while you're standing in line at a Disney park? You'll have to bring it with you from home. In an effort to keep chewed gum from being stuck all over the parks, none of the shops in any Disney theme park sells gum. Supposedly this innovation came from Walt Disney himself, who wanted to make sure that his guests could enjoy their visits without getting gum stuck to their shoes.

7 Nikita Khrushchev

Disneyland as a battleground for the Cold War? Believe it or not, that's exactly what it became in 1959. That year, Soviet leader Nikita Khrushchev spent 11 days visiting the United States. He spent one day of the trip in Los Angeles, and the fierce orator wanted to see Disneyland. However, the LAPD and the rest of Khrushchev's security detail were worried about his safety during such a trip, so they nixed the idea.

Khrushchev accepted the news with characteristic poise, which is to say he exploded. He ranted, "And I say, I would very much like to go and see Disneyland. But then, we cannot guarantee your security, they say. Then what must I do? Commit suicide? What is it? Is there an epidemic of cholera there or something? Or have gangsters taken hold of the place that can destroy me?"

Hundreds of feral cats patrol Disneyland at night, which really helps keep the non-cartoon mice in check.

10 LISTS
TO LIGHTEN THE MOOD AT THE E.R.

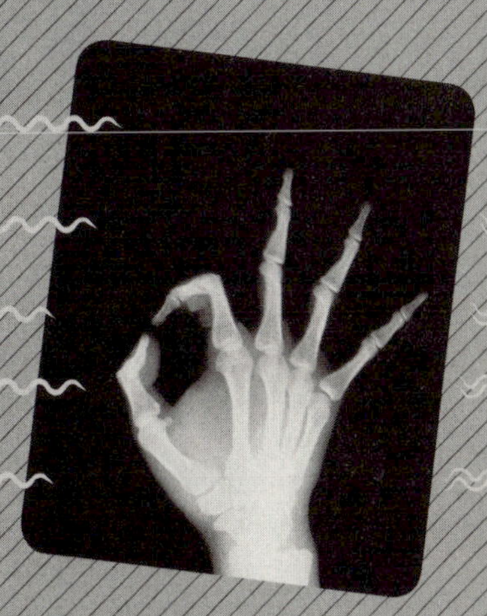

COULD BE WORSE!

THREE DEFUNCT DISEASES YOU DEFINITELY DON'T HAVE

Virgin's Disease

You know you've got it if: You're suffering from green skin, menstrual cessation, and lethargy.
Victims: In 1554, doctors determined the green monster was targeting virgin girls with the disease they labeled "chlorosis." Later, various physicians reported that the condition was a direct result of women either being undersexed or, in the case of university girls, overeducated.
Treatment: Many believed the cure to ending virgin's disease was as simple as ending virginity. In a letter to a worried father, one physician suggested that he arrange for his daughter to get pregnant as soon as possible. His rationale? "If they conceive, they recover." Amazingly, chlorosis didn't disappear from medical textbooks until the 1930s. These days, doctors recognize the symptoms as part of anemia and prescribe iron supplements instead of sex.

A flu outbreak canceled the 1919 Stanley Cup when too many players were ill to play past the fifth game.

2 Visceroptosis, or "Organ Drooping"

You know you've got it if: You think you're sick. If you suffer from occasional headaches, poor sleep, or even if you don't have any real symptoms, organ drooping is probably to blame.

Victims: People with poor posture, women who had multiple pregnancies, and—above all—girls who wore excessively tight corsets. Visceroptosis was defined as the downward displacement of inner organs within the abdominal cavity. Testing was simple: If a doctor placed light pressure on patients' abdomens and it made them feel better, organ drooping was taking place.

Treatment: Although organs can cause problems if they get repositioned in the body, the diagnosis was basically a way for surgeons to make money. Organ drooping was such a common diagnosis at the end of the 19th century that specialized surgery clinics popped up across the country to "treat" it. But the popularity of visceroptosis ended with World War I, when surgeons had real problems to fix.

3 The English Sweat

You know you've got it if: You're experiencing fever, aches, exhaustion, and, of course, sweating through your shirt. Worse still, people were said to die within 24 hours of contracting the symptoms.

Victims: Strangely, only people living in England. Outbreaks of the sweating sickness broke out in the summer months in 1485, 1508, 1517, 1528, and 1551. Only once did an outbreak make it beyond England's borders.

The real cause: Poor hygiene. Although scientists still aren't sure exactly what caused "the sweating sickness," they believe it might have been a flu-type virus spread by filth or rodents. One monarch had a unique prevention technique: King Henry VIII was so scared of contracting the sweat that he moved around the country from manor to manor trying to outrun it.

911 became the emergency phone number in 1968. The number's merits: It's easy to remember, and 9 and 1 are so far apart on a rotary phone it's hard to misdial it.

. . . AND TWO REAL DISEASES YOU MIGHT HAVE

1 Lovesickness

You know you've got it if: You're listening to a lot of country music. In addition to some unrequited love, you also may experience loss of appetite, trouble sleeping, and an irregular pulse, among other things.

What it isn't: One ancient medieval writer claimed the illness could cause the body of a jilted lover to fill with black bile. Also, an Islamic philosopher said lovesick men could turn into werewolves.

What it could be: Roman Emperor Commodus' personal physician, Claudius Galenus, first officially diagnosed lovesickness as a medical disease in the 2nd century CE. Although that classification eventually fell out of favor, recent brain-imaging studies have shown that people who are madly in love exhibit neurological patterns similar to OCD sufferers.

2 Lycanthropy

You know you've got it if: You're pretty sure you're turning into a werewolf.
What it isn't: Supernatural forces turning you into a werewolf, even if you *are* lovesick.
What it could be: While people don't actually shapeshift into wolves, lycanthropy is a very rare, very real psychological syndrome in which patients develop the delusion that they can transform into animals. Mythological explanations of lycanthropy involved demonic possession or nasty wolf bites, but today's psychologists believe that clinical lycanthropy is the result of schizophrenia or mood disorders.

Obsessive nose picking is called rhinotillexomania.

YOU'RE WELCOME

FIVE THINGS YOU ALWAYS WANTED TO KNOW ABOUT YOUR BODY

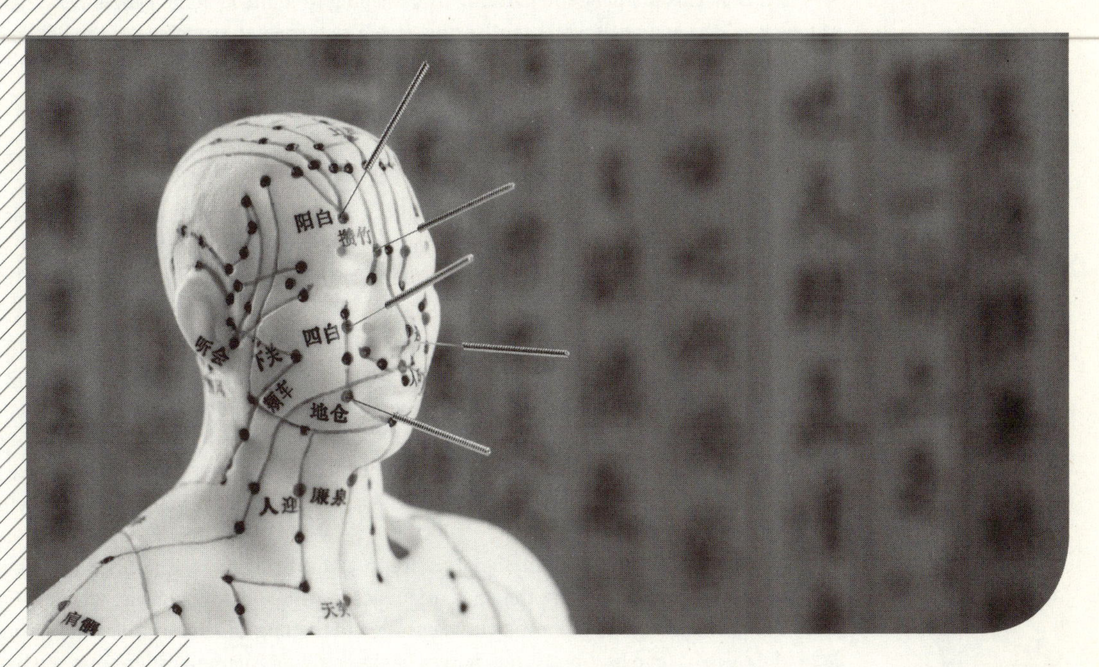

❶ What Causes an "Ice Cream Headache"?

The "ice cream headache" phenomenon is the enemy of many of us who try to cool off too quickly in the summertime. When an icy cold substance comes into contact with the roof of the mouth, the nerves there send a signal to the body to dilate the blood vessels in the head in an attempt to keep the brain warm and functioning. This rush of blood is what causes the short-lived "headache," which in most cases is intense for only about 15 seconds and then dissipates completely in under a minute.

These "brain freezes" only occur in one-third of the population. So if you're part of that unlucky minority, but you've just gotta have that Rocky Road, here are a couple of tricks to ease the pain. Keeping the treats away from the roof of your mouth will prevent the headache. And if you can wait until winter, studies have shown that the reaction doesn't occur when the weather is cold.

The "G" in "G-Spot" stands for Ernst Gräfenberg, a German-born doctor who studied female sexual physiology.

What Makes the "Popping" Sound When People Crack Their Knuckles?

Put simply, it's the sound of gas bubbles bursting. In joints like that of the finger, bones are connected by a "joint capsule" made of tissue, lubricated with what is known as synovial fluid. When the joint is stretched too far, gas bubbles are formed in the fluid, and these bubbles burst with an audible "pop." Once a particular joint is cracked, it takes a short period of time (about half an hour) for the gas to dissolve back into synovial fluid, making popping possible again.

But is all this popping and cracking as damaging to joints as some people would have you believe? Yes and no. It won't cause arthritis, but the repeated stretching of the ligaments at the ends of the joints may damage them. The hand may become swollen, and the ability to grip may be affected.

❸ Why Do Our Body's Limbs Occasionally "Fall Asleep"?

When the body is resting in an unusual position, it can block the nerve signals and/or blood flow to a particular part of the body. Since nerves require both signals from the brain and nutrients from the blood in order to function properly, this interruption is what causes a limb to "fall asleep."

When these brain communications are first interrupted, a prickly numbness known as paresthesia occurs, alerting our body to the notion that something is wrong and coercing us to readjust our position. In fact, we're lucky we get this feeling because prolonged obstruction can result in permanent damage. This condition is commonly referred to as "Saturday night palsy," a limb injury that can occur when an overly intoxicated person falls into a deep sleep and cannot wake when the body sends these signals. (Enough to make you think twice before tipping the bottle, huh?)

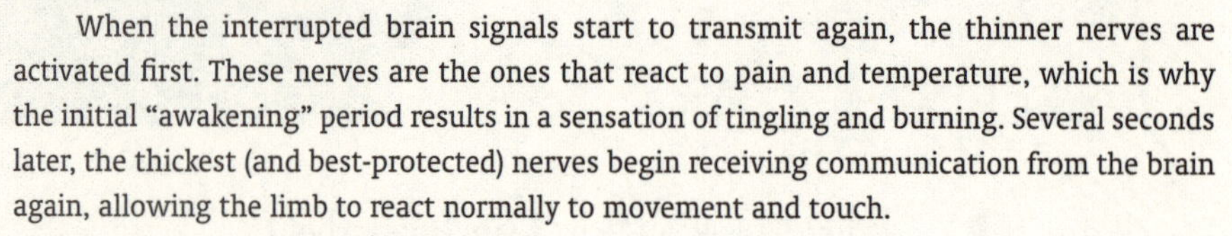

When the interrupted brain signals start to transmit again, the thinner nerves are activated first. These nerves are the ones that react to pain and temperature, which is why the initial "awakening" period results in a sensation of tingling and burning. Several seconds later, the thickest (and best-protected) nerves begin receiving communication from the brain again, allowing the limb to react normally to movement and touch.

❹ Why Are Some People Left-Handed?

Approximately 10 percent of the U.S. population is left-handed; on that much experts can agree. But what makes some people favor one hand over another? Part of the difficulty in pinpointing a cause is that scientists can hardly agree on what constitutes a left-handed person. For example, if you write with your left hand, you may consider yourself a "southpaw." But for a researcher to consider you to be a true lefty, he'd run you through a battery of at least 10 tasks, such as striking a match and shooting a marble, to determine which of your hands was truly dominant. While today left-handedness is accepted as a fact of life, like brown eyes or red hair, there was a time when teachers and parents forced their children to

In 1965, President Lyndon Johnson issued the first Medicare card. The recipient? Harry Truman.

use their right hands, which explains why fewer people 65 and older (only six percent) write with their left hands.

Scientists agree that handedness is something that's determined in the womb, but exactly how it's determined is still being debated. Most authorities agree that the level of testosterone in the pregnant mother has something to do with it, which is perhaps why the majority of lefties are male. For years, it was thought that handedness was genetically predisposed, but this theory had a (left-handed) monkey wrench thrown in it when a study was done on twins. Identical twins have identical genes, yet of the lefty twins studied, only 76 percent were both left-handed. Additionally, studies have shown that when both parents are left-handed, there is only a 26 percent chance that their children will be left-handed.

5 Why Do People Snore?

Snores emanate from the very back of the throat. When you sleep, the muscles around the upper part of the throat's airway relax, making the surrounding tissue soft and flaccid, diminishing the space in the airway. When breathing becomes difficult, the forcing of air to and from the lungs causes these tissues to "flap" as they separate to allow the air to pass. This creates a buzzing, vibrating, and often highly annoying sound.

About half the adult population occasionally snores, and 25 percent regularly do so. Because aging causes the deterioration of muscles and obesity causes fatty deposits to appear around them, these factors result in an increased amount of snoring for those affected.

Touch Your Glabella! Seven Body Parts You Can't Name

1. **Niddick:** the technical term for the nape of your neck
2. **Philtrum:** the groove between your nose and your upper lip. It was once believed that the philtrum was one of the most erogenous parts of the body.
3. **Gluteal fold:** the crease where your buttocks meets your legs
4. **Glabella:** the flat, smooth plane between your eyebrows
5. **Nasion:** the internal indent where your nose meets your forehead, right under your glabella
6. **Ginglymus:** the inner elbow
7. **Acnestis:** the part of your back you can't reach

The boundary between your lips and the skin of your face is called the vermilion border.

HANDS OFF!

THE MISSING BODY PARTS OF SEVEN FAMOUS PEOPLE

Dan Sickles' Leg

During the Battle of Gettysburg, Major General Daniel Sickles was sitting on his horse when a Confederate cannonball hit his right leg and almost tore the thing off. Though the general was reportedly so unfazed by the event that he smoked a cigar en route to the medical tent, Sickles' leg had to be amputated. The nonplussed Sickles saved his detached limb and later donated it to the National Museum of Health and Medicine in Washington, D.C. He even found a convenient use for the extremity: picking up chicks. Apparently, Sickles would bring lady-friends to the museum when he wanted to impress them with his tales of bravery. The rest of Sickles was buried at Arlington National Cemetery after his passing in 1914.

Dr. Thomas Harvey removed Einstein's brain during his autopsy. For 30 years, Harvey kept it in two mason jars in his Wichita home.

"Stonewall" Jackson's Arm

Confederate General Thomas "Stonewall" Jackson got his nickname by sitting astride his horse "like a stone wall" while bullets whizzed around him during the Civil War. But that kind of bravery (or foolhardiness) didn't serve him well. During the Battle of Chancellorsville, Jackson was accidentally shot in the arm by one of his own men. Said arm had to be amputated, and afterward, it was buried in the nearby Virginia town of Ellwood. Only eight days later, Stonewall was stone-cold dead of pneumonia. The rest of his body is resting in peace in Lexington, Va.

3 Saint Francis Xavier's Hand

Francis Xavier was a saint with a few too many fans. In the early 16th century, the Spanish missionary was sent to Asia by the king of Portugal to convert as many souls to Christianity as possible. Turns out, he was pretty good at the job. Francis Xavier became wildly popular, and after his death in 1552, so did his relics. In fact, demand out-fueled supply. Throughout several years and multiple exhumations, his body was whittled away. Today, half his left hand is in Cochin, India, while the other half is in Malacca, Malaysia. One of his arms resides in Rome, and various other cities lay claim to his internal organs. The leftovers? They went to Goa, India.

Saint Catherine of Siena's Finger

Ever think you're going to pieces? Saint Catherine feels your pain. After the holy woman died in 1380, her body became an object of veneration. Pilgrims believed touching her miraculously un-rotted flesh could heal illnesses and bring them closer to God, so they flocked to visit the body from all over Europe. Eventually, the Catholic Church laid Catherine to rest—part of her, at least. Before she was buried, one of her followers removed a finger (along with a few teeth and other various and sundry body parts). Meanwhile, Pope Urban VI got a similar idea and took her head. Today, both finger and head are on display at San Domenico Church in Siena, Italy. The rest of her is beneath the main altar at Santa Maria Sopra Minerva Church in Rome.

Horses and rats don't have gallbladders.

5 Napoleon's Penis

Exiled emperor Napoleon Bonaparte died on May 5, 1821. The following day, doctors conducted an autopsy, which was reportedly witnessed by many people, including a priest named Ange Vignali. Though the body was said to be largely intact at the time of the undertaking, it seems the priest took home a souvenir. In 1916, Vignali's heirs sold a collection of Napoleonic artifacts, including what they claim to be the emperor's penis. While no one knows for sure if it really is Napoleon's, uh, manhood, people have paid good money for the penis. Currently, it's in the possession of an American urologist.

6 Oliver Cromwell's Head

Oliver Cromwell, the straitlaced Puritan who usurped the English throne, wasn't exactly a wild man. His head, however, was sometimes the life of the party. Cromwell died in 1658, but two years later, the reinstated English monarchy exhumed, tried, and hanged his body, then dumped it in an unmarked grave. In addition, as a warning to would-be killers, his head was placed on a pike in Westminster Hall, where it remained for 20 years. After a subsequent stint in a small museum, it was sold in 1814 to a man named Josiah Henry Wilkinson (perhaps looking to parade it around as an exceptionally gruesome icebreaker at parties). Such was the ironic afterlife of the Puritan until 1960, when his head was finally laid to rest in a chapel in Cambridge.

7 Sarah Bernhardt's Leg

Ever tell an actor to "break a leg"? Be careful what you wish for. In 1905, the Divine Sarah injured her knee performing the last scene of the play "La Tosca." Sadly, the injury never healed. By 1915, gangrene had set in and the leg had to be amputated. Afterward, she continued to perform, sticking to roles that allowed her to remain seated. According to legend, circus mastermind P. T. Barnum offered Bernhardt a hefty chunk of change for the amputated leg, but she turned him down. The true whereabouts of the appendage remains a mystery.

The Smithsonian has a "Presidential Hair Case" with the hair of 13 presidents, including Washington, John Adams, Jefferson, Madison, and Pierce.

FOOTNOTES

SEVEN LITTLE-KNOWN TOE FACTS

1 They Can Replace Your Fingers

Since fingers and toes are both digits, they should be interchangeable, right? Well, in toe-to-hand surgery, toes can be used to replace missing fingers. The method was first used on humans in 1975 and is now widely used. Not every finger can be replaced, but often the big toe can be used for a missing thumb.

2 You Can't Serve without One (Even Though You Probably Could)

To enlist in the army, you have to have all 10 of your toes intact—army regulations state that they will reject anyone for "current absence of a foot or any portion thereof." But as it turns out, it shouldn't be an issue, since a missing toe doesn't cause much damage. In fact, doctors have been able to design special shoes for toe amputees that will correct minor step problems.

3 Anna Wintour Thinks They're Sexy

Vogue Editor-in-chief Anna Wintour once listed the "unwritten dress code" of the Voguette, which included "toe-cleavage shoes, sans stockings." She's not the only one to preach the sexiness of toe cleavage; opening the vamp of a shoe is well regarded as a chic move in the fashion world. Frederick's of Hollywood says showing toes can be sexually suggestive. However, the key is moderation. Manolo Blahnik warns that the secret is only showing the first two cracks.

4 Stalin's Were Webbed

Joseph Stalin suffered from syndactyly, where multiple digits are fused together. In other words, the toes (or fingers) are webbed. There's no evidence that the condition causes any problems, nor does it improve swimming ability, as might be expected. Other famous sufferers of syndactyly are Dan Aykroyd and Ashton Kutcher.

5 The Egyptians Knew How to Replace Them

Even though a missing toe won't cause significant problems, that doesn't mean they shouldn't be replaced. In fact, toe prosthetics could date back as far as 3,000 years. Explorers found a mummy in Egypt with a leather-and-wood contraption that is believed to be a prosthetic toe. The "Cairo Toe" dates to between 1069 BCE and 664 BCE and predates the earliest known prosthetic by at least 700 years.

Argentina's future First Lady Eva Peron sang aspirin radio jingles in the mid-1940s. Before long, the country became the world's largest per-capita consumer of the drug.

You Can Wrestle Them

If you can arm wrestle and thumb wrestle, doesn't it just make sense that you can toe wrestle. Since 1993, England has played host to the World Toe Wrestling Championship, a contest with too many toe puns to repeat. Contestants simply lock toes in a ring, then try to push each other out in a three-round toe-down. (Get it?) Despite the cult popularity of the sport, the International Olympic Committee rejected it when organizers applied for inclusion in the Games. The sport's superstar is Paul Beech, who nicknamed himself "The Toeminator."

People Have Had as Many as 13

The Guinness Book of World Records currently lists a tie for the most number of fingers and toes at 25. Two Indian boys, Pranamya Menaria and Devendra Harne, each have 12 fingers and 13 toes thanks to polydactylism, a congenital condition that results in extra digits. Polydactilism occurs in about one in every 500 births and can be treated. Marilyn Monroe is rumored to have been born with an extra toe on her left foot, but the proof is iffy.

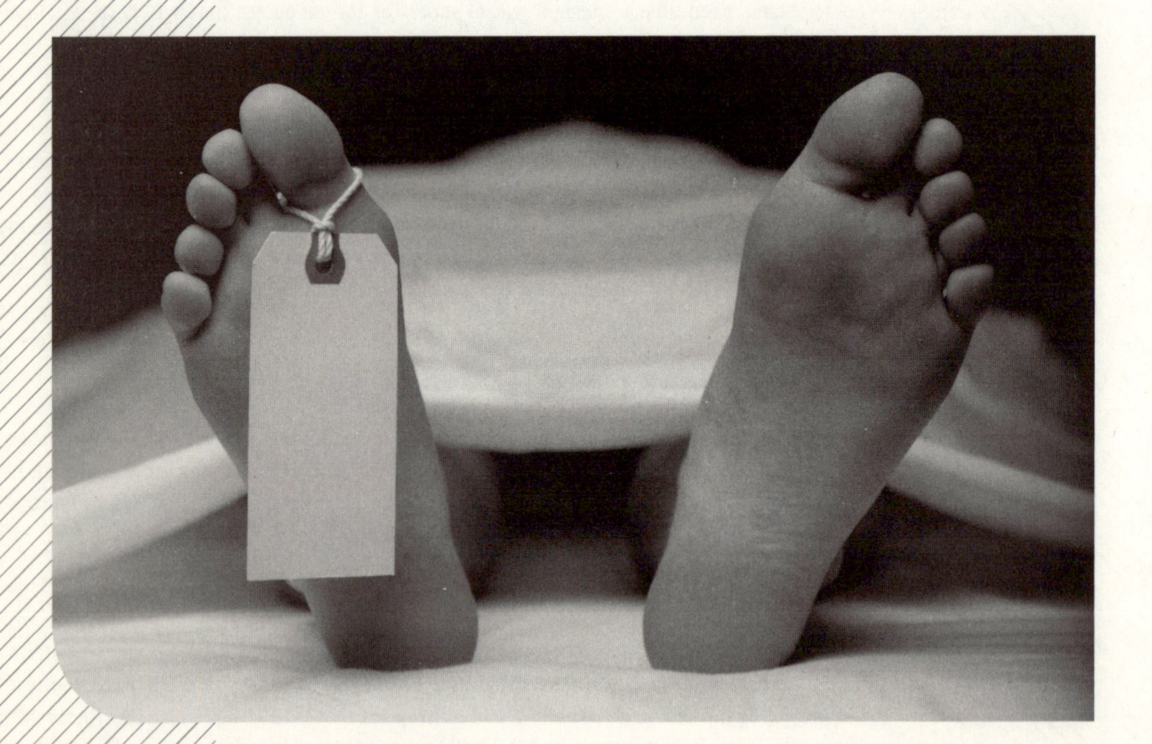

In the Brothers Grimm version of *Cinderella*, one stepsister cuts off her own toes in an attempt to make the shoe fit.

OUR HEADS ARE SPINNING

FIVE MIND-BOGGLING PSYCHIATRIC TREATMENTS

1 Insulin-Coma Therapy

The insulin-coma therapy trend began in 1927. Viennese physician Manfred Sakel accidentally gave one of his diabetic patients an insulin overdose, and it sent her into a coma. But what could have been a major medical faux pas turned into a triumph. The woman, a drug addict, woke up and declared her morphine craving gone. Then Sakel (who really isn't earning our trust here) made the same mistake with another patient, who also woke up claiming to be cured.

Before long, Sakel was intentionally testing the therapy with other patients and reporting a 90 percent recovery rate, particularly among schizophrenics. Strangely, however, Sakel's treatment successes remain a mystery. Presumably, a big dose of insulin causes blood sugar levels to plummet, which starves the brain of food and sends the patient into a coma. But why this unconscious state would help psychiatric patients is anyone's guess. Regardless, the popularity of insulin therapy faded, mainly because it was dangerous. Slipping into a coma is no walk in the park, and between one and two percent of treated patients died as a result.

2 Rotational Therapy

Charles Darwin's grandfather Erasmus Darwin was a physician, philosopher, and scientist, but he wasn't particularly adept at any of the three. Consequently, his ideas weren't always taken seriously. Of course, this could be because he liked to record them in bad poetic verse (sample: "By immutable immortal laws / Impress'd in Nature by the great first cause, / Say, Muse! How rose from elemental strife / Organic forms, and kindled into life").

It could also be because his theories were a bit far-fetched, such as his spinning-couch treatment. Darwin's logic was that sleep could cure disease and that spinning around really fast was a great way to induce the slumber. Nobody paid much attention to it at first, but later, American physician Benjamin Rush adapted the treatment for psychiatric purposes. He believed that spinning would reduce brain congestion and, in turn, cure mental illness. He was wrong. Instead, Rush just ended up with dizzy patients who were still crazy. These days, rotating chairs are limited to the study of vertigo and space sickness.

3 Mesmerism

Much like Yoda, Austrian physician Franz Mesmer (1734–1815) believed that an invisible force pervaded everything in existence, and that disruptions in this force caused pain and suffering. But Mesmer's ideas would have been of little use to Luke Skywalker. His basic

Gasoline was once sold in small bottles as a cure for lice.

theory was that the gravity of the moon affected the body's fluids in much the same way it caused ocean tides, and that some diseases accordingly waxed and waned with the phases of the moon. The dilemma, then, was to uncover what could be done about gravity's pernicious effects.

Mesmer's solution: Use magnets. After all, gravity and magnetism were both about objects being attracted to each other. Thus, placing magnets on certain areas of a patient's body might be able to counteract the disruptive influence of the moon's gravity and restore the normal flow of bodily fluids. Surprisingly, many patients praised the treatment as a miracle cure, but the medical community dismissed it as superstitious hooey and chalked up his treatment successes to the placebo effect. Mesmer and his theories were ultimately discredited, but he still left his mark. Today, he's considered the father of modern hypnosis because of his inadvertent discovery of the power of suggestion, and his name lives on in the English word "mesmerize."

4 Malaria Therapy

Ah, if only we were talking about a therapy for malaria. Instead, this is malaria as therapy—specifically, as a treatment for syphilis. There was no cure for the STD until the early 1900s, when Viennese neurologist Wagner von Jauregg got the idea to treat syphilis sufferers with malaria-infected blood. Predictably, these patients would develop the disease, which would cause an extremely high fever that would kill the syphilis bacteria. Once that happened, they were given the malaria drug quinine, cured, and then sent home happy and healthy.

The treatment did have its share of side effects—that nasty sustained high fever, for one—but it worked, and it was a whole lot better than dying. In fact, Von Jauregg won the Nobel Prize for malaria therapy, and the treatment remained in use until the development of penicillin came along and gave doctors a better, safer way to cure the STD.

5 Chemically Induced Seizures

Nobody ever said doctors had flawless logic. A good example: seizure therapy. Hungarian pathologist Ladislas von Meduna pioneered the idea. He reasoned that, because schizophrenia was rare in epileptics, and because epileptics seemed blissfully happy after seizures, then giving schizophrenics seizures would make them calmer. In order to do this, von Meduna tested numerous seizure-inducing drugs (including such fun candidates as strychnine, caffeine, and absinthe) before settling on metrazol, a chemical that stimulates the circulatory and respiratory systems. And although he claimed the treatment cured the majority of his patients, opponents argued that the method was dangerous and poorly understood.

To this day, no one is quite clear on why seizures can help ease some schizophrenic symptoms, but many scientists believe the convulsions release chemicals otherwise lacking in patients' brains. Ultimately, the side effects (including fractured bones and memory loss) turned away both doctors and patients.

Poppy seeds really can make you test positive for heroin. Prisons often forbid foods with poppy so drug-test results aren't blamed on bagels.

I'M STILL HERE!

SIX DEATHS THAT WERE GREATLY EXAGGERATED

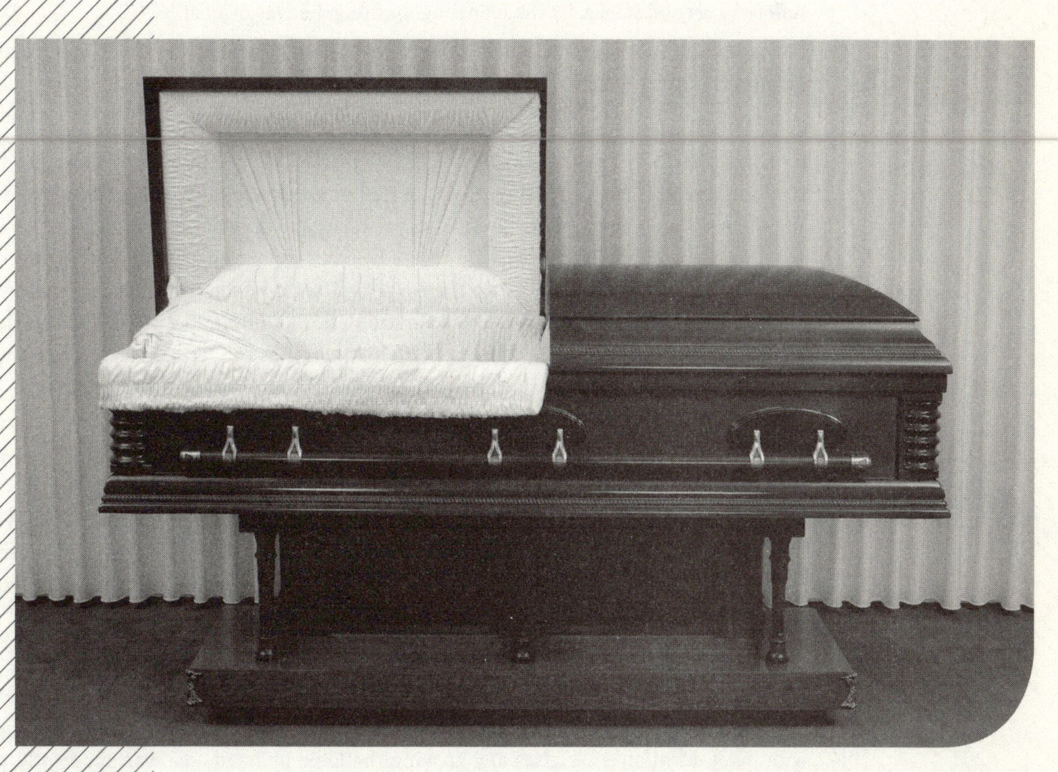

The Cast of *Cannibal Holocaust*

Still one of the most controversial films ever made, the 1980 Italian exploitation-fest *Cannibal Holocaust* depicted such realistic and horrifying violence that Italian authorities believed it was an actual snuff film. Ten days after its release, authorities confiscated prints of *Holocaust* and arrested its director on suspicion of murder. Not helping matters much was the fact that the film's cast had signed agreements saying they would lay low for a full year after the film's release, fueling rumors that they were, in fact, slaughtered for the camera. Finally facing life in prison, the director voided his actors' "no-media" contracts so they could come forward to clear his name.

When P. T. Barnum fell gravely ill at age 81, he convinced the *New York Sun* to publish his obituary in advance so he could see it in print.

Unflattering Obituary Kills Marcus Garvey

A stroke incapacitated black nationalist leader Marcus Garvey in 1940. Rumors began to circulate that he had died, and before Garvey could quell them, the *Chicago Defender* ran an obituary that described him as a man who died "broke, alone and unpopular." When Garvey read the unflattering passage he let out a loud moan and collapsed to the floor, where he suffered a second stroke. By the following morning, he was dead at 53.

❸ Things to Do in Texas When You're Dead

In his 16-year career, major league relief pitcher, Bill Henry, played for the Boston Red Sox and the Cincinnati Reds, rang up 46 wins, and even pitched in the 1961 World Series. In August 2007 the Lakeland, Fla., *Ledger* reported that Henry had passed away at the ripe old age of 83, and the Associated Press picked up the story for national distribution.

Bill Henry didn't live in Lakeland where he had supposedly died, though. He lived (and still lives) in Deer Park, Texas. Once the *Ledger* got wind of the truth, a very strange story came to light: Another man named Bill Henry, a salesman from Florida, had stolen the ball player's identity and spent 20 years passing himself off as the retired major league pitcher. The fake Henry, who was 83 when he died, had fooled everybody—including his wife—who later said, "I was married to somebody that maybe I didn't know."

How did the impostor explain the incorrect birthday listed on his baseball card? "A printing error." The "fake" Bill Henry even gave lectures twice a year at a Florida college entitled "Baseball, Humor and Society." After the matter was cleared up, however, the real Bill Henry harbored no ill feelings. "I just hoped maybe it helped him in his [sales] career," he said.

❹ The Not-Quite-Canonized Thomas à Kempis

Well-known medieval author-monk Thomas à Kempis, it is said, was accidentally buried alive in 1471. A most decidedly low-temperature dude in life—he spent most of his time engaged with quiet devotional exercises and copying the Bible by hand—he was apparently not so cool under pressure when it came to death. When authorities exhumed his body some time later, they found scratch marks on the underside of the coffin and splinters of wood under his fingernails. As if it wasn't bad enough to be buried alive, when the Church discovered the tragedy, they promptly shut down efforts to canonize Kempis as a saint. Their reasoning? "Surely no aspiring saint, finding himself so close to meeting his maker, would fight death in this way!" Talk about adding insult to being buried alive . . .

Hans Christian Andersen was so paranoid about being buried alive that each night he left a note next to his bed that read, "I only appear to be dead."

5 Samuel Taylor Coleridge

In 1816, the writer heard his name mentioned in a hotel by a man reading a coroner's report in the newspaper, who remarked that "it was very extraordinary that Coleridge the poet should have hanged himself just after the success of his play, but he was always a strange mad fellow." Coleridge replied: "Indeed, sir, it is a most extraordinary thing that he should have hanged himself, be the subject of an inquest, and yet that he should at this moment be speaking to you." (Now that's what I call a killer comeback!) Turns out a man had been found hanging from a tree in Hyde Park—an apparent suicide—and the only identification he had was the name "S. T. Coleridge" written on the inside of the collar of his shirt. Coleridge thought the shirt had probably been stolen from him.

6 Hiroo Onoda, the Soldier Who Wouldn't Die

A Japanese soldier stationed in the Philippines during World War II, Hiroo Onoda was presumed dead after the Allies recaptured the country in 1945. But he and a few comrades had fled into the jungle to hide, and for 29 years, that's where he stayed. Unwilling to believe that the war had ended, he and his scrappy fellows continued to launch mini-attacks against Filipino citizens that killed dozens over the years. In 1959, he was declared legally dead in Japan, and by 1972, when the last of his compatriots were killed in gunfights with local forces, Onoda was finally alone.

Onoda stayed for two more years, until the Japanese government found his old commanding officer from the war—he had become a bookseller many years before—who was flown to the jungle, where he informed Onoda of the defeat of Japan in WWII and ordered him to lay down his arms. Lieutenant Onoda emerged from the jungle 29 years after the end of World War II, and accepted the commanding officer's order of surrender in his dress uniform and sword, with his Arisaka Type 99 rifle still in operating condition, 500 rounds of ammunition, and several hand grenades. Onoda later wrote a book about his experiences and started a nature camp for kids designed to teach them survival skills.

Bengay cream is named for Jules Bengué, the French pharmacist who developed the pain reliever in the late 1800s.

THE CURIOUS DEATHS OF 10 INTERESTING PEOPLE

Despite our best efforts, Death, in all its myriad and weird forms, is constantly lurking around the corner. But who knew a toothpick could be so dangerous? Or that one's trademark scarf, draped so dramatically around your neck, could be conspiring to kill you?

1. King Adolf Frederick of Sweden ate himself to death in 1771: His last meal included lobster, caviar, cabbage, smoked herring, and Champagne, followed by 14 servings of his favorite dessert, semla in hot milk.

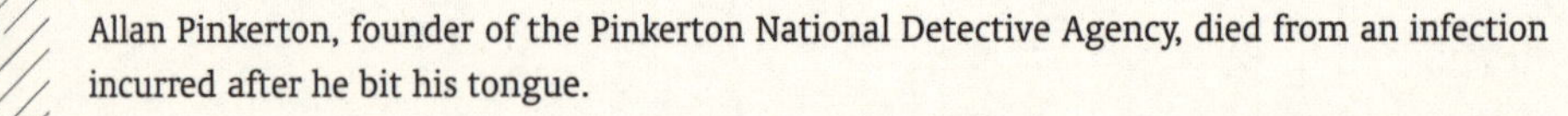

2. Allan Pinkerton, founder of the Pinkerton National Detective Agency, died from an infection incurred after he bit his tongue.

Hospitals adopted blue and green scrubs in the mid 20th century. The darker hues didn't have to be replaced as often as whites and were easier on doctors' eyes.

Jack Daniel, purveyor of fine whiskey, died from an infection sustained after kicking his safe and busting his toe.

Isadora Duncan, an early-20th-century modern dancer, was killed by her trademark scarf while riding in a convertible car. The long scarf blew back and wrapped around a tire axel, breaking Duncan's neck.

Tennessee Williams, longtime alcoholic and author of some of the most enduringly bleak plays of the 20th century, choked on an eyedropper bottle cap in 1983.

Sherwood Anderson, author of *Winesburg, Ohio*, contracted and died from peritonitis—an infection of the lining of the stomach—after he swallowed part of a toothpick.

According to legend, Aeschylus, the Greek playwright, died after an eagle dropped a tortoise on his head. The tortoise reportedly lived.

8

Attila the Hun allegedly died of a nosebleed on his wedding night; he passed out drunk and drowned in his own blood.

9

Draco, the Greek lawmaker whose stringent legal code gave rise to the word "draconian," died somewhere in the 7th century BCE, supposedly after a particularly masterful speech. He suffocated under the mounds of hats and cloaks thrown upon him by admiring Greeks as a show of appreciation.

10

King Henry I died in 1135 of food poisoning after gorging on lampreys, a parasitic eel-like marine animal popular in British cuisine during the Middle Ages. Because he died while in France, his remains were sewn into the hide of a bull and shipped back to England for burial.

Henry Wadsworth Longfellow's wife died when a dropped match ignited her enormous hoop skirt.

WORK IN PEACE

FIVE THINGS YOUR BODY CAN DO AFTER YOU DIE

1 Get Married

Death is no obstacle when it comes to love in China. That's because ghost marriage—the practice of setting up deceased relatives with suitable spouses, dead or alive—is still an option.

Ghost marriage first appeared in Chinese legends 2,000 years ago, and it's been a staple of the culture ever since. At times, it was a way for spinsters to gain social acceptance after death. At other times, the ceremony honored dead sons by giving them living brides. In both cases, the marriages served a religious function by making the deceased happier in the afterlife.

While the practice of matchmaking for the dead waned during China's Cultural Revolution in the late 1960s, officials report that ghost marriages are back on the rise. Today, the goal is often to give a deceased bachelor a wife—preferably one who has recently been laid to rest. But in a nation where men outnumber women in death as well as in life, the shortage of corpse brides has led to murder. In 2007, there were two widely reported cases of rural men killing prostitutes, housekeepers, and mentally ill women in order to sell their bodies as ghost wives. Worse, these crimes pay. According to the *Washington Post* and the *London Times*, one undertaker buys women's bodies for more than $2,000 and sells them to prospective "in-laws" for nearly $5,000.

Frank Hayes is the only dead jockey to win a race. In 1923, he had a heart attack mid-ride, but his body stayed upright through the finish.

2 Unwind with a Few Friends

Today, most of us think of mummies as rare and valuable artifacts, but to the ancient Egyptians, they were as common as iPhones. So, where have all those mummies gone? Basically, they've been used up. Europeans and Middle Easterners spent centuries raiding ancient Egyptian tombs and turning the bandaged bodies into cheap commodities. For instance, mummy-based panaceas were once popular as quack medicine. In the 16th century, French King Francis I took a daily pinch of mummy to build strength, sort of like a particularly offensive multivitamin. Other mummies, mainly those of animals, became kindling in homes and steam engines. Meanwhile, human mummies frequently fell victim to Victorian social events. During the late 19th century, it was popular for wealthy families to host mummy-unwrapping parties, where the desecration of the dead was followed by cocktails and hors d'oeuvres.

3 Fuel a City

Cremating a body uses up a lot of energy—and a lot of nonrenewable resources. So how do you give Grandma the send-off she wanted and protect the planet at the same time? Multitask. Some European crematoriums have figured out a way to replace conventional boilers by harnessing the heat produced in their fires, which can reach temperatures in excess of 1,832°F. In fact, starting in 1997, the Swedish city of Helsingborg used local crematoriums to supply 10 percent of the heat for its homes.

4 Stand Trial

In 897 CE, Pope Stephen VI accused former Pope Formosus of perjury and violation of church canon. The problem was that Pope Formosus had died nine months earlier. Stephen worked around this little detail by exhuming the dead pope's body, dressing it in full papal regalia, and putting it on trial. He then proceeded to serve as chief prosecutor as he angrily cross-examined the corpse. The spectacle was about as ludicrous as you'd imagine. In fact, Pope Stephen appeared so thoroughly insane that a group of concerned citizens launched a successful assassination plot against him. The next year, one of Pope Stephen's successors reversed Formosus' conviction, ordering his body reburied with full honors.

When the mummy of Ramses II was sent to France in 1976, it was issued a passport. Ramses' occupation? "King (deceased)."

5 Become a Soviet Tourist Attraction

Russian revolutionary Vladimir Lenin wanted to be buried in his family plot. But when Lenin died in 1924, Joseph Stalin insisted on putting his corpse on public display in Red Square, creating a secular, Communist relic. Consequently, an organization called the Research Institute for Biological Structures was formed to keep Lenin's body from decay. The Institute was no joke, as some of the Soviet Union's most brilliant minds spent more than 25 years working and living on-site to perfect the Soviet system of corpse preservation. Scientists today still use their method, which involves a carefully controlled climate, a twice-weekly regimen of dusting and lubrication, and semiannual dips in a secret blend of 11 herbs and chemicals.

Unlike bodies, however, fame can't last forever. The popularity of the tomb is dwindling, and the Russian government is now considering giving Lenin the burial he always wanted.

Lenin spent more than three years with a bullet lodged in his neck from a failed 1918 assassination attempt. Doctors finally cut out the slug in April 1922.

TOMB RAIDING 101

FOUR FAMOUS GRAVE ROBBERIES

1 Stealing the Tramp

Silent-era funnyman Charlie Chaplin died on Christmas day in 1977 and was buried soon after in Switzerland. But in March 1978, grave robbers swiped Chaplin's body and phoned in a ransom demand of £400,000. The thieves' plan seemed so perfect until Chaplin's widow, Lady Oona Chaplin, refused to pay the sum because "Charlie would have thought it rather ridiculous."

In an attempt to nab the crooks, the local police set up false payoff meetings, but these proved fruitless when the robbers didn't show. However, both police and the suspects were persistent, so the two parties continued to communicate in the hopes of resolving the standoff.

In May, the police were expecting another call from the robbers, so they tapped the Chaplins' phone. In an extraordinary display of coordination, they also assigned officers to watch as many as 200 phone booths throughout the area. When the call from the robbers came in, police traced it back to the originating booth and arrested two men, Roman Wardas and Gantscho Ganev, both auto mechanics. The men led police to Chaplin's remains, which were buried in a cornfield about 10 miles from the graveyard.

Wardas received a four-year stint for masterminding the scam, while Ganev, seen only as a muscleman, got off easy with an 18-month suspended sentence. As for Chaplin, he was reburied in the same burial plot, but this time his coffin was surrounded by thick concrete to prevent anyone else from disturbing his slumber.

2 The Slinking Memorial

In the early hours of November 7, 1876, a group of four counterfeiters broke into Oak Ridge Cemetery in Springfield, Ill., with the intention of stealing Abraham Lincoln's body from his sarcophagus. They planned to take the body, hide it in the sand dunes of northern Indiana, and hold it for $200,000 ransom, plus demand the release of one of their gang from prison.

A police informant in the crew foiled the plan, though. When the men broke into the cemetery that night, police and Secret Service agents (who were only charged with investigating counterfeiters at the time, not guarding the body of the President) were waiting for them. The crooks initially got away, only to be arrested a few days later.

After the attempted robbery, Lincoln's remains were reburied in the same mausoleum at Oak Ridge, but instead of being inside the sarcophagus, they were secretly hidden in a shallow grave in the basement of the tomb—a fact that was known only to a handful of people for decades. There the body stayed until 1901, when eldest son Robert Todd Lincoln had his father's remains placed inside a steel cage, lowered 10 feet into the ground, and covered in concrete for safekeeping.

Charlie Chaplin once entered a Charlie Chaplin look-alike contest in a theater in San Francisco. He lost.

Elvis Almost Left the Building

In August 1977, just two weeks after The King's death, police received word from an informant that a group planned to steal Elvis Presley's 900-pound, steel-lined, copper-plated coffin and hold his remains for ransom.

Thanks to this heads-up a police task force monitored Presley's grave and caught three men—Raymond Green, Eugene Nelson, and Ronnie Adkins—snooping around the mausoleum. Just how the men planned on getting through the two concrete slabs and solid sheet of marble that covered the coffin is unknown, since they weren't carrying any tools or explosives. The Memphis police felt like something about the situation didn't add up, so until further evidence about the plot could be uncovered, they charged the men with criminal trespassing and kept them in jail.

As the investigation continued, it became apparent that the story the informant had told police was full of holes. He said the men were going to be paid $40,000 each by a mysterious criminal mastermind who planned to ransom the body for $10 million. But he couldn't tell police how the men intended to get their reward or how to contact this shadowy kingpin once the deed had been done. With no actual crime being committed (other than the men being in the cemetery after dark), and the evidence against the men being so weak, all charges were eventually dropped.

As a result of the almost, kinda, sorta attempted grave robbery, the Presley estate requested permission to move the bodies of Elvis and his mother to Graceland where they could be monitored 24-hours a day by staff security and closed-circuit TV cameras. Of course they're still at Graceland and have become one of the estate's main attractions.

4 Stay On the Line. Police Will Be with You Shortly.

Soon after his death in 2001, the body of Enrico Cuccia, a powerful bank president often considered the father of Italian capitalism, disappeared from its vault. A loyal housekeeper who visited the grave on a weekly basis to clean up around the tomb discovered the foul play.

The family received a ransom demand for the equivalent of $3.5 million to be deposited by Mediobanca—the bank Cuccia had controlled for more than 50 years—into a numbered Swiss account. When the ransom was not immediately paid, a man called Mediobanca to set up the transfer of funds, but was placed on hold under the pretense that the bank president was on the other line. This gave the police time to trace the call back to a small village near Turin, Italy, and found Giampaolo Pesce, a steelworker, still holding the phone.

Caught red-handed, Pesce led authorities to a barn where Cuccia's coffin had been hidden under some straw.

In 1977, a tree branch that fell to the ground during Elvis' funeral sold at auction for $748.

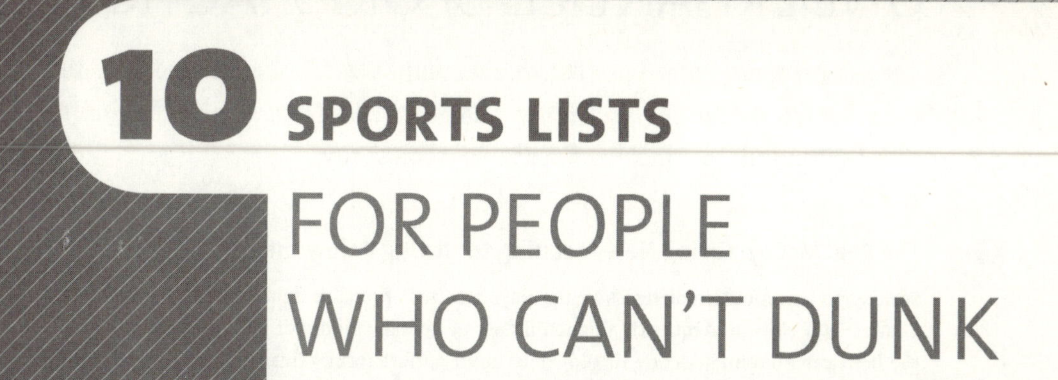

10 SPORTS LISTS FOR PEOPLE WHO CAN'T DUNK

IF YOU'RE NOT CHEATING, YOU'RE NOT TRYING

SIX GLORIOUSLY UNDERHANDED SPORTS TACTICS

Cheating is unimaginative, brutish, and plain crass. Underhandedness, on the other hand, requires a certain mustache-twirling panache—a boldness that beguiles us, no matter what the rule books say!

1 The Real McCoy: Giving New Meaning to Hitting Below the Belt

Seeking to psych-out challengers in the days leading up to big fights, Hall of Fame boxer Charles "Kid" McCoy frequently feigned illness or spread rumors of an injury. Then, when the bout came around, McCoy would show up in perfect form. (This supposedly prompted reporters to wonder whether they'd be seeing "the real McCoy" in the ring.)

McCoy's lowest blow came when he fought a deaf mute in 1893, though. Toward the end of the fourth round, McCoy simply dropped his gloves and walked back to his corner as though the bell had sounded. When the deaf fighter turned to do the same, McCoy ran over and knocked him out.

2 Red Auerbach: The Host from Hell

Coach Arnold "Red" Auerbach, the cigar-chomping mastermind behind the great Boston Celtics teams of the 1950s and 1960s, wasn't one to let any advantage go unused. Auerbach knew his home stadium inside and out and manipulated it to create one of the greatest home court advantages in the history of sport. To foster a feeling of alienation among opposing players, he would assign visiting teams a different locker room in the Boston Garden each time they came to town. To foster a feeling of nausea, he reportedly made sure at least one toilet in the visitors' quarters was stopped up and overflowing. And finally, to foster a feeling of "it's so hot I'm gonna die," he contrived to have the building's boilers stoked and steaming right before tip-off and again at halftime.

3 The Spanish Paralympic Basketball Team: Playing Dumb

The grand champions of sport ethics obliteration have to be the members of the 2000 Spanish Paralympic basketball team. How low could they go? After the team snagged a gold medal, it was revealed that 10 of the team's12 players had never been tested, and were, in fact, not mentally challenged.

At the 1904 Olympics in St. Louis, American gymnast George Eyser grabbed one bronze, two silvers, and three gold medals—all while competing with a wooden leg.

4 Eddie Stanky and the Stanky Maneuver

One of the all-time greats at probing the limits of sports rule books was second baseman, Eddie "The Brat" Stanky. The best evidence of Stanky's creative rule interpretations came in 1950, when baseball commissioner Ford Frick had to forbid Stanky from using what had become known as the "Stanky Maneuver," a dubious defensive tactic in which he took advantage of his position behind the pitcher by "jumping up and down while waving wildly in an attempt to distract opposing batters."

5 Jason Grimsley and the Ol' Bat-and-Switch

When suspicions arose that Cleveland Indians slugger Albert Belle had been corking his bat, it made for a cloak-and-dagger spy-fest. Tipped off about Belle's bat during a 1994 game against the White Sox, umpires confiscated it and took it to a locker room for later investigation. Knowing Belle's bat was doctored, and not wanting to lose their best offensive player to a suspension, the Indians dispatched pitcher Jason Grimsley to sneak into the room and switch out the bat for a legal model. Grimsley climbed through about 10 feet of ducts and a false ceiling to pull the switch. The plan might have worked, if only he hadn't replaced it with the autographed bat of teammate Paul Sorrento. The caper was quickly discovered, and Belle soon found himself suspended.

6 Gene Bossard and the Field of Streams

For groundskeeper Gene Bossard, lending the Chicago White Sox an underhanded hand was the family business. Gene managed the turf at Comiskey Park from 1940 to 1983, and when he stepped down, his son Roger took over operations. Together, the Bossards were known for doctoring and dampening the diamond to give the Sox a true home field advantage. In fact, opposing teams took to calling the infield "Bossard's Swamp," because Gene kept it watered down to benefit the Sox's sinkerball pitchers and to slow opposing baserunners.

Bossard's most infamous trick, however, seems to be inventing the "frozen baseball." Perhaps Roger Bossard explained the phenomenon best: "In the bowels of the old stadium my dad had an old room where the humidifier was constantly going. By leaving the balls in that room for 10 to 14 days, they became a quarter- to a half-ounce heavier." The Sox' manager during the frozen ball era in the late 1960s? Number 4 on this list, Eddie Stanky.

Even though he never played college football, the Dallas Cowboys drafted Pat Riley as a receiver in the 11th round of the 1967 NFL draft.

TO RENEW YOUR FAITH

FIVE CHEER-WORTHY ACTS OF SPORTSMANSHIP

1 Lutz Long Gives Jesse Owens a Boost

At the 1936 Summer Olympics in Berlin, German long jumper Lutz Long set an Olympic record during the preliminary round to qualify for the finals. American Jesse Owens fouled on his first two attempts and faced disqualification if he fouled again. Before Owens made his final attempt, Long, a German, advised him to adjust his takeoff point—to several inches behind the foul line—to ensure that he would advance to the next round.

Owens heeded Long's advice, qualified for the finals, and set a new world record to win the gold medal. Long took the silver. "It took a lot of courage for him to befriend me in front of Hitler," Owens later said. "You can melt down all the medals and cups I have and they wouldn't be a plating on the 24-carat friendship that I felt for Lutz Long at that moment." Long was killed in World War II, but his family has remained in contact with Owens' family ever since.

Elmo Wright of the Kansas City Chiefs became the first NFL player to perform an end zone dance after catching a TD pass in 1973.

2 Jack Nicklaus Picks Up the British Ryder Cup Team

The 1969 Ryder Cup at the Royal Birkdale Club in Southport, England, was tied as the final pair, the United States' Jack Nicklaus and England's Tony Jacklin, teed off on the 18th hole. Nicklaus, who was playing in his first Ryder Cup, sank his four-foot par putt. Before Jacklin could address his two-foot par putt to tie, Nicklaus reached down and picked up his opponent's ball marker. It was a sporting gesture by Nicklaus, who didn't want to put Jacklin through the pressure of making the "gimme" before thousands of British fans.

By conceding the putt, Nicklaus ensured that the competition would end in a tie for the first time in its 42-year history. "I don't think you would have missed that putt . . . but in these circumstances I would never give you the opportunity," Nicklaus told Jacklin.

3 John Landy's Scrape with Greatness

Australian John Landy made history when he became the second man to break 4 minutes in the mile, 46 days after Roger Bannister became the first in 1954. Landy is revered in Australia, where he served as the 26th Governor of Victoria, in part because of the mile race he ran at the 1956 Australian national championships.

During the third lap, 19-year-old Ron Clarke, who would go on to set 17 world records during his career, tripped and fell. Landy, who was trailing close behind, leaped over Clarke and accidentally scraped his rival's arm with his spikes in the process. Landy stopped running to make sure that Clarke wasn't badly hurt before resuming his chase of the pack that had charged ahead. To the amazement of everyone in the crowd, Landy came from behind to finish first in a time of 4 minutes, 4 seconds.

Fifty years after the fact, Landy reflected on the astonishing race. "I reacted on the spur of the moment," he said. "You do things like an embedded impulse. You don't ask why."

4 High School QB Gives Back Record

Star quarterback Nate Haasis of Southeast High School in Springfield, Ill., set the Central State Eight Conference record for career passing yards in the final minute of a loss to Cahokia High in 2003. Haasis was excited to set the record, but he thought it was strange that Cahokia's defenders backed 20 yards off the line of scrimmage and made no attempt to defend or tackle the receiver who caught his record-setting pass. The next day, the local newspaper reported that the coaches had made a deal to allow Haasis to set the record, a story both coaches confirmed. "I had my guys put their arms in their jerseys so they couldn't tackle," Cahokia's coach later said.

The first organized intercollegiate sporting event in history was the University Boat Race between Oxford and Cambridge in 1829.

Three days after the game, Haasis decided to write a letter to the director of the conference, requesting that his final pass be omitted from the conference record book. "I would like to preserve the integrity and sportsmanship of a great conference for future athletes," Haasis wrote. His request was granted.

5 Soccer Player Uses Hands, Draws Cheers

During a 2000 English Premier League match between West Ham and Everton, Paolo Di Canio displayed an act of sportsmanship that, as one reporter wrote, "will live longer than the forgettable game it accompanied." With the match tied in extra time, Everton goalkeeper Paul Gerrard injured his knee. While Gerrard lay writhing in pain, another West Ham player sent a cross toward Di Canio, who waited in front of the wide-open goal. Rather than receiving the pass and scoring the go-ahead goal, Di Canio caught the ball to allow Gerrard to be treated. His unselfish act drew a standing ovation and earned him FIFA's Fair Play Award in 2001.

Tom Selleck was the honorary captain of the 1984 Men's Olympic Volleyball team.

BEYOND STEROIDS

10 UNBELIEVABLE PERFORMANCE-ENHANCING SUBSTANCES

Looking for a boost on the playing field? Here are 10 performance enhancers that put the "dope" in doping.

1 Strychnine

Better known as a deadly poison, strychnine was thought to be the ticket to success for Thomas Hicks, an American runner in the 1904 Olympics marathon, who mixed it with brandy and egg whites. Sure, he needed emergency medical treatment at the finish line and nearly died, but he also won the gold medal.

Asses' Hooves

The hooves of the Abyssinian ass were taken in powder form by ancient Egyptian athletes, who boiled them in oil and spiked them with rose hips to mask that funky ass-hoof flavor.

Sweet, Sweet Ether

Sugar cubes soaked in ether helped cyclists get through 144-hour races in the 1870s. When ether wasn't enough, coaches added nitroglycerine and cocaine. Don't be too horrified, though; they also tossed in some peppermint for flavor.

4 Baking Soda

Some modern-day swimmers and runners swear by "soda-doping," which means ingesting baking soda to shave precious seconds off their times. By increasing blood pH, baking soda may reduce the acid produced by athletes' muscles, allowing them to keep on trucking. Soda-doping also causes diarrhea, which could conceivably help keep the competition out of the pool.

From the late 1920s through the 1950s the NFL periodically used a white football in night games to help players see the ball in poor light.

5 Ultraviolet Radiation

In the 1930s, Russian scientists found that peppering athletes with UV rays improved their speed in the 100-meter dash. German researchers in the 1940s found similar improvement in swimming times after a spell under the sunlamp.

6 Guinea Pig Sperm

In the late 1800s, physiologist Charles Brown-Séquard injected himself with his *liquide testiculair* and later testified that the extract of dog and guinea-pig testes made him stronger and had even "lengthened the arc of [his] urine." While he likely overstated the effectiveness of his own distillation, his discovery was a predecessor of hormone-based performance-enhancers.

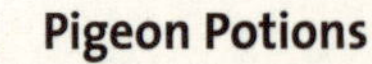

7 Pigeon Potions

The sport of pigeon racing has been rocked by doping scandals: While some handlers resort to anabolic steroids, others use drugs that prevent molting to maximize a bird's training season. Birds have also been doped with laxatives to encourage them to drop their payloads before racing, thereby reducing the birds' weight.

8 Magic Mushrooms

It wasn't just Norse berserkers who favored magic mushrooms to enhance performance: Olympians in the third century also counted on hallucinogenic 'shrooms to speed them to the finish line.

9 Arsenic

In the regions of Styria and Tyrol, Austrian lumberjacks kept the ax swinging with megadoses of arsenic, which would have made a nifty addition to the Monty Python song. Styrians also took arsenic to pep them up before long mountain hikes and as a digestive aid.

10 Human Hearts

The ancient Aztecs swore by 'em to boost both military and athletic prowess.

At the 1912 Olympic Games in Stockholm, General George S. Patton finished fifth in the first-ever modern pentathlon.

CEREAL SNUBS

FIVE SPORTS HEROES YOU WON'T FIND ON A WHEATIES BOX

1 Sumo Wrestling: Akebono Tarō

The only slim thing about sumo wrestling is the chance of becoming a *yokozuna*, or grand champion. Throughout the centuries, only 69 men have done it. Before Hawaii's Chad Rowan stomped into the ring, no foreigner had ever held the honor. Of course, improbable things can happen when you stand 6'8" and weigh more than 500 lbs.—gigantic even by sumo standards. After abandoning a college basketball scholarship due to arguments with his coaches, Rowan threw himself into sumo.

In 1988, he went to Japan with only a single set of clothes and a limited knowledge of Japanese. But Rowan wasn't there to chitchat. Within a year, the quick study had learned how to use his towering height to make devastating thrusts at opponents' throats. That March, he made his professional debut as Akebono—"dawn" in Japanese—an ironic moniker for a man who could block out the sun.

At the Wife Carrying World Championships in Sonkajärvi, Finland, first prize is the wife's weight in beer.

As Rowan's victories piled up and his Japanese improved, he won more and more fans. His jovial demeanor didn't hurt, either. In January 1993, Akebono was promoted to *yokozuna*—a title he held until retirement. By the time he was ready to hang up his belt in 2001, he'd racked up 566 wins and 11 division championships.

2 Bullfighting: Sidney Franklin

In 1922, Sidney Franklin was just an artist from Brooklyn who'd moved to Mexico City after an argument with his father. One day, he decided to take a break from painting to see his first bullfight. Franklin immediately fell in love with the sport—particularly the crowd's reverence for the fighters. When he told his Mexican friends that he was surprised by the absence of American matadors, they replied that Americans didn't have the guts to step into the arena. The ribbing irritated Franklin so much that he embarked on a quixotic mission to become a legendary bullfighter.

In need of a trainer, Franklin brashly solicited the services of renowned Mexican matador Rodolfo Gaona. The request was basically the equivalent of asking Peyton Manning for free football lessons, but shockingly, Gaona accepted.

Franklin's fearlessness didn't translate into instant success. During his first fight in 1923, he fell down twice before killing the bull. Within five years, however, he was thrilling Mexican crowds. But the victories weren't enough for Franklin. Looking for bigger challenges, he set out to conquer the motherland of toreadors—Spain. Franklin's gutsy performances in Spanish arenas earned him throngs of fans, along with several gorings. They also earned him the friendship of bullfighting aficionado Ernest Hemingway. The author would later immortalize Franklin's technique and bravery in *Death in the Afternoon*, saying Franklin's life story was "better than any picaresque novel you ever read."

3 Billiards: Willie Mosconi

It's hard to believe that billiards world champion Willie Mosconi learned to play pool by hitting potatoes with a broomstick. It's even harder to believe that his parents, who ran a pool hall in Philadelphia, forbade him from playing because they wanted him to pursue a career in vaudeville. Luckily for them, the obstinate Mosconi taught himself late at night with the only implements at his disposal.

In no time, Mosconi became a cue-wielding child prodigy. His talents supported his family during the Great Depression, and Mosconi went on to win 15 world championships during his career. Impressively, he still holds the world record for running balls without a miss, sinking 526 consecutive balls in a 1954 exhibition.

Of course, Paul Newman might argue that Willie Mosconi's greatest accomplishment was teaching him to play pool. Allegedly, Newman had never played before filming *The Hustler*. After taking intense pool-shark lessons from Mosconi, however, Newman was nominated for an Academy Award for best actor in 1962.

Ohio State University has produced so many Olympians that if the school were a country, its 77 medals would rank 31st overall.

4 Polo: Sue Sally Hale

It's tough to imagine anyone taking the title of "Sports' Best Cross-Dresser" away from Hale. Hale, who received her first horse at the age of 3, was determined to play polo, even though Southern California's thriving early 1950s polo scene forbade women from the field. So when she was old enough to play, Hale simply dressed as a man. Before each tournament, she would don a baggy shirt, stuff her hair under her helmet, and draw on a mustache with mascara. Playing under the name A. Jones, she competed with such ferocity that one commentator claimed Hale "could ride a horse like a Comanche and hit a ball like a Mack truck."

After each match, she would transform back into Sue Sally Hale, then go carousing with her teammates, who were happy to play along. For the next two decades, Hale maintained the ruse while campaigning fiercely to get the United States Polo Association to change its policies. The association relented in 1972, and Hale finally received a membership card, along with the freedom to play under her real name.

5 Cricket: John Barton King

Cricketers in the United States may be traditionally associated with wealthy men of leisure, but the top player ever produced this side of the pond was a middle-class baseball fan from Philly named Bart King. What made King so great was his ability to dominate both as a bowler and a batsman—the equivalent of being a top-notch pitcher and slugger in baseball. As a bowler, King created a pitch he called "the angler," which dipped and swerved in a way that confounded batsmen. As a batter, he was one of the top scorers in North American history.

The gregarious King was also beloved for spreading tall tales about himself. Perhaps his most famous story came from a 1901 match against a team from Trenton, New Jersey. As the legend goes, King was about to bowl to the Trenton team captain when the batter started to talk trash. Remembering a stunt he'd seen in a baseball game, King ordered the rest of his team off the field. He reasoned that he wouldn't need anyone around to catch the ball, because he was about to strike out the loudmouthed batter. The cocky move proved effective. King fired off his angler, and the befuddled Trenton captain didn't stand a chance.

When he first came to the United States, 7'7" Manute Bol's passport listed him as 5'2". "They measured me while I was sitting down," he said.

NOT READY FOR THE BIG LEAGUES

FIVE MAJOR SCANDALS FROM FIVE MINOR SPORTS

1 It's a Sprint, Not a Marathon

Cuban-American runner Rosie Ruiz didn't just win the 1980 Boston Marathon, she set a new record with a time of 2:31:56. However, on closer inspection, it turned out Ruiz probably hadn't run the whole race. Or even most of it. No one saw Ruiz plodding along in the early going, and she somehow shaved over 25 minutes off her impressively fast time in the 1979 New York Marathon only six months earlier, further raising eyebrows.

It turned out that maybe the New York Marathon time wasn't completely legit, either; a freelance photographer came forward with the revelation that she had definitely been with Ruiz on the subway during the race.

Soon, a narrative formed: It seemed that Ruiz had cheated in the New York Marathon, and cheated so well she'd posted an outstanding sub-three-hour time and qualified for Boston, a major achievement for any marathon runner. Her boss was so excited about this triumph

British golfers won the first 16 editions of the U.S. Open.

that he offered to pay her expenses to run in Boston. At this point, Ruiz was probably too embarrassed to fess up to her earlier misdeed, so she went to Boston and waited at Kenmore Square, around a mile from the finish line, jumped into the race, and sprinted to the finish.

Most observers don't think Ruiz was trying to win, just post a respectable time, but she jumped in too early and set a new record. Marathon officials stripped Ruiz of the title after interviewing her and finding she knew very little about the course's landmarks or distance-running jargon, but she still maintains that she finished both races fair and square. As such, Ruiz has never returned her first place medal.

Fishy Results

In 2005 angler Paul Tormanen of Lee's Summit, Missouri, was a rising star on the competitive bass fishing circuit, often grabbing his limit of fish within an hour of a contest opening. His career seemed to really be taking off, at least until he was arrested in Louisiana for felony contest fraud. Tormanen admitted a fairly basic scheme for winning some big-money bass tournaments; he'd catch his fish beforehand, take them out on the lake, and tie them to stumps. He used his tethered fish in an ill-fated attempt to win the 2005 CITGO Bassmaster Central Open, a competition that offered its champion a $10,000 cash prize and a new fishing boat.

Unfortunately for Tormanen, another competitor found one of his ringer fish during a practice round and secretly marked it with the help of fish and wildlife officials. When Tormanen weighed in with his catch, authorities caught onto his fraud. The incident earned Tormanen a lifetime ban from B.A.S.S. competitions, and he received a suspended sentence of six months, a fine, 120 hours of community service, and two years of probation

3 Badminton: Suddenly Dangerous

Although badminton is usually just played at picnics and in backyards in the United States, it's a very popular competitive sport throughout Asia. On July 28, 1988, it even turned deadly in India. Syed Modi, a popular figure who had won the national championship eight times as well as a gold at the Commonwealth Games and a title at the 1983 and 1984 Austrian International, was gunned down by a group of men as he left a practice session in Lucknow.

The murder became the talk of the Indian press, with speculation raging that the murder was masterminded by one of Modi's friends, who was also rumored to be the lover of Modi's wife. Other members of the press contended this arrest was a red herring perpetrated by prime minister Rajiv Gandhi. Although police later brought charges against seven people, only one was convicted, and the motive for the slaying is still unclear.

In the 1940s and 1950s, the grass surface on most miniature golf courses was actually goat hair that had been dyed green.

Camp Barbed Wire

Rugby union is a major passion in South Africa, and the national team, known as the Springboks, wanted to win the 2003 Rugby World Cup so much that they went a bit overboard in their preparations. When the roster for the event was named in September 2003, coach Rudolph Straeuli decided to send the squad to a police camp in the South African bush. The activity, known as Kamp Staaldraad, or "Camp Barbed Wire," would bring the players together as a team.

This excursion was no corporate team-building retreat, though. It was a bit more brutal: Players were allegedly forced at gunpoint into a freezing lake to pump up rugby balls, then dumped naked into a foxhole where icy water was poured on their heads as they sang the national anthem. Other reports included the news that the players were forced to crawl naked across gravel and kill chickens.

When the South African media got wind of this training exercise it became a full-blown scandal that cost Straeuli his job and earned the contempt of most fans. Even worse, the fracas demoralized the Springboks, who couldn't make it past New Zealand in the quarterfinals.

5 Drug Racing

Critics occasionally like to poke fun at NASCAR's alleged roots of Southern moonshining and bootlegging, but the now-defunct IMSA GT race circuit was rife with real smuggling during its brief life as an alternative racing league in North America. From at least 1975 to 1986 a handful of top drivers on the tour paid for their racing teams not just by selling sponsorships, but by operating a massive drug-smuggling cartel.

How big was their outfit? When the drivers were caught, it was estimated that they'd imported and distributed over 300 tons of Colombian marijuana over the course of eight years. Several drivers, including John Paul, Sr., John Paul, Jr., Randy Lanier, and the Whittington brothers were convicted in connection with the ring. Former 12 Hours of Sebring winner John Paul, Sr. was the alleged mastermind of the operation; he received a 25-year federal sentence for charges that included shooting a federal witness. Pundits noted that the initials IMSA must have stood for "International Marijuana Smugglers Association."

The 1976 Winter Olympics were awarded to Denver, but Colorado voters rejected the public financing of the Games. Innsbruck hosted instead.

INSULT TO INJURY

13 UNLUCKY AND UNUSUAL SPORTS INJURIES

1. Sacramento Kings small forward Lionel Simmons missed two games in the 1991 season because he had tendonitis in his right wrist and forearm from playing too much Game Boy.

2. In 1994, journeyman knuckleballer Steve Sparks missed out on a chance to make his first big-league roster when he dislocated his shoulder during spring training with the Milwaukee Brewers. He tried to rip a phone book while imitating a group of motivational speakers who had visited the team.

3. NHL goalie Glenn Healy enjoyed a long career, and he also enjoyed playing the bagpipes. While playing for the Maple Leafs in 2000, Healy needed stitches after slicing himself while repairing an antique set of pipes.

4. Point guard Muggsy Bogues once missed the second half of a game after he became dizzy from inhaling ointment fumes while receiving treatment for a sore muscle at halftime.

5. Hall of Fame offensive tackle Turk Edwards suffered a career-ending injury in 1940 during the pregame coin toss. When Edwards turned to return to the Washington Redskins' sideline, he caught his cleats on the turf, which wrecked his fragile knee and forced him into retirement.

Go Cornjerkers! 10 Unbelievable High School Mascots

1. Avon (CT) Old Farms Winged Beavers
2. Blooming Prairie (MN) Awesome Blossoms
3. Williamsport (PA) Millionaires
4. Ridgefield (WA) Spudders
5. Hoopeston Area (IL) Cornjerkers
6. New Berlin (IL) Pretzels
7. Watersmeet (MI) Nimrods
8. Frankfort (IN) Hot Dogs
9. New Braunfels (TX) Unicorns
10. Poca (WV) Dots

Wrestler Andre the Giant supposedly once drank 117 bottles of German beer in a single sitting.

6. After scoring a goal in 2004, Servette midfielder Paulo Diogo decided to celebrate by jumping into the crowd. His wedding ring had other plans, though, and caught on a fence. The force ripped off the top of Diogo's finger, and to add insult to injury, the refs booked him for excessive celebration.

7. Actually, it might just be a good idea to take off all of your rings when you hit the field or court. Atlanta Braves closer Cecil Upshaw missed the entire 1970 season when his ring got caught on an awning as he tried to demonstrate his slam dunk technique.

8. Boston Red Sox pitcher Clarence Blethen didn't have a long career, but he did have a fine set of false teeth. According to legend, Blethen liked to carry his false teeth in his back pocket when he played, which was a decent enough plan until he had to make a hard takeout slide at second base during the 1923 season. When Blethen slid, his false chompers supposedly bit him on the rear, which led to a bloody mess that forced him out of the game.

9. Journeyman pitcher Greg A. Harris once missed two starts for the Texas Rangers after he inflamed his elbow by spending an entire game flicking sunflower seeds at a friend who was sitting nearby.

10. Norwegian soccer defender Svein Grøndalen went for a jog as part of his training for an international match during the 1970s . . . and ran headlong into a moose. The injury forced him to withdraw from the match.

11. Chicago Cubs outfielder Jose Cardenal missed a game in 1974 because his eyelid was "stuck open," which prevented him from blinking. Although he eventually overcame this ailment, it didn't help Cardenal's reputation as a player who liked to use suspicious injuries to get out of games; two seasons earlier he had missed a game because crickets in his hotel room had kept him up all night, leaving him exhausted.

12. In May 2002 Baltimore Orioles left fielder Marty Cordova fell asleep in a tanning bed and got a wicked sunburn on his face. His doctor ordered him to stay out of direct sunlight, which meant Cordova had to hide out in the clubhouse during day games until his face healed.

13. Stuttgart Kickers soccer player Sascha Bender once suffered a facial injury after being punched. The assailant, teammate Christian Okpala, said Bender "permanently provoked me by farting all the time."

In 1957, Phillies centerfielder Richie Ashburn fouled off a pitch and broke a fan's nose. He fouled off the next pitch, too—and hit the same fan.

NO BIG DEAL

SIX OF BASEBALL'S STRANGEST TRADES

Harry Chiti for Harry Chiti

The mysterious "players to be named later" is often a key piece in baseball trades, and there are usually restrictions on what players can actually be named later. The strangest player named later, though, was Harry Chiti. At the beginning of the 1962 season, the Cleveland Indians dealt catcher Chiti to the New York Mets for cash and a player to be named later. In June, the two teams decided on the player: Harry Chiti. Essentially, Chiti was traded for himself and cash, bringing new meaning to the old term "rent-a-player."

Johnny Jones for a Turkey

Depression-era owner Joe Engel of the minor league Chattanooga Lookouts was a publicity hound with a knack for headline-grabbing promotions like giving away a house during a game. Perhaps his most unusual stunt came when he traded shortstop Johnny Jones to Charlotte, though. In return, Engel received a 25-pound turkey, which he prepared for the media. After trying the turkey, Engel declared that Charlotte had won the trade because the turkey was tough. Maybe if that turkey had been juicy, Chattanooga would have come out ahead.

The Pittsburgh Steelers drafted Johnny Unitas in 1955. They cut him after the last game of the 1955 preseason. Whoops.

Oddly enough, that isn't baseball's only player-for-food trade. In 1998, the Pacific Suns traded Ken Krahenbuhl to the Greensville Bluesmen for a player, cash, and 10 pounds of Mississippi catfish.

3 Marilyn Peterson for Susan Kekich

Yankees pitchers Mike Kekich and Fritz Peterson were friends and, in the swinging '70s had even engaged in some innocent wife-swapping. But in 1973, they took it a step further, literally switching wives. The ladies moved into their new partner's house, bringing the kids and even the dogs with them. Baseball commissioner Bowie Kuhn said he was appalled when he found out, but the Yankees had a lighter look at it. General manager Lee MacPhail joked, "We may have to call off Family Day."

4 Joe Gordon for Jimmy Dykes

As GM of various baseball teams, Frank Lane built up a reputation as being quick to the trigger on any trade, earning him the nicknames "Frantic Frank" and "Trader Lane." His most unusual trade came in the middle of the 1960 season when he swapped his manager, Joe Gordon, to Detroit in exchange for Tigers manager Jimmy Dykes. The move was mostly a stunt—Lane allegedly wanted to trade the entire team to the Chicago White Sox, but was stopped by the commissioner of baseball—and didn't help either team. The Indians finished in fourth place in the American League, two places ahead of the Tigers.

5 John Odom for 10 Bats

In 2008, the Calgary Vipers acquired pitcher John Odom only to learn that a minor blotch on his criminal record would keep him from entering Canada. Odom's inability to make it through Canadian immigration rendered him pretty much worthless to the Vipers, so the team set out to trade him for whatever it could get. Their final haul: a set of 10 maple bats from the Laredo Broncos. This wasn't the Vipers' first foray into trading players for merchandise, either: The team had previously tried to trade a player for seats for their new stadium.

6 Dave Winfield for Dinner

Hall of Famers like Dave Winfield usually fetch astronomical prices on the trade market, but in the waning years of his career, Winfield was traded for dinner. The Minnesota Twins dealt Winfield to the Indians for a player to be named later at the trade deadline in 1994. The players' strike ended the season before Winfield could ever take the field for the Indians, though, and a player was never named. To settle the trade, executives from Minnesota and Cleveland decided to go out for dinner, and the Indians picked up the check.

Lee Corso and Burt Reynolds were roommates while playing football for Florida State in the mid-1950s.

"SHOW ME THE JELL-O!"

SIX ATHLETE CONTRACT CLAUSES ONLY AN AGENT COULD IMAGINE

1 Charlie Kerfeld's Tasty Bonus

After a spectacular rookie season in 1986, the rotund reliever needed a new contract. Kerfeld asked for $110,037.37 to match his number 37 jersey to pitch in 1987. On top of that, he received 37 boxes of orange Jell-O in the deal. The Astros would soon regret this delicious bonus, though; Kerfeld, who was famously caught eating ribs in the dugout that season, battled weight and injury problems and was sent down to the minors.

2 Support for Rollie Fingers' 'stache

Former A's owner Charlie Finley never thought of a gimmick he wouldn't try, including a mechanical rabbit that delivered fresh balls to the umpire and hiring 13-year-old MC Hammer as his "Executive V.P." In 1972, Finley offered his players cash to grow a mustache by Father's Day, thereby giving birth to reliever Fingers' trademark handlebar 'stache. The A's went on to win the World Series that season, and Fingers' contract for 1973 contained a $300 bonus for growing the mustache as well as $100 for the purchase of mustache wax.

3 Roy Oswalt's Big Ride

Before Oswalt made a start in the 2005 National League Championship Series, Astros owner Drayton McLane promised to make the ace's dreams come true if he won, specifically his life goal of bulldozer ownership. After Oswalt dominated the Cardinals to send Houston to its first-ever World Series, McLane came through with a Caterpillar D6N XL. Since Major League Baseball requires high-dollar gifts be disclosed, Oswalt signed an addendum to his contract, a "bulldozer clause," authorizing the club to give him his new toy.

4 George Brett Becomes a Landlord

The contract extension George Brett signed with the Kansas City Royals in 1984 must have been one of the stranger deals in MLB history. The club agreed to give Brett the bat he used in the infamous 1983 "Pine Tar Game" as part of the pact, but that wasn't the only odd perk. The contract also gave him part ownership of an apartment complex in Memphis.

At the time, Avron Fogelman co-owned the Royals. Fogelman had made his fortune as a lawyer and real estate baron in Memphis, so when the team needed a little extra incentive to get Brett to sign, they offered the third baseman a piece of one of Fogelman's developments. Brett's agent/brother Bobby negotiated the deal; he referred to the 1,100-unit apartment

Alabama football coach Bear Bryant was once asked to contribute $10 to help pay for a sportswriter's funeral. According to legend, he said, "Here's a twenty, bury two."

complex as "a nice little kicker." Brett received a guaranteed cash flow of $1 million from the development and retained the right to sell his 10-percent stake to the Royals for $2 million.

Fogelman's Royals used the trick a couple more times when they signed reliever Dan Quisenberry and outfielder Willie Wilson to similar deals that gave them stakes in the 700-unit Stewart's Ferry apartment development in Nashville. (Quiz got 24.5 percent of the development, while Wilson got 9.5 percent.) Fogelman later told the *New York Times* that he might have given up too much in the hastily negotiated deals.

5 Curt Schilling Stays Skinny

By the end of his storied career outspoken hurler Curt Schilling had started to get a bit doughy. When the Boston Red Sox re-signed him to a one-year deal with $8 million before the 2008 season, it included a clause in which Schilling could pick up an extra $2 million if he made weight at six random weigh-ins over the course of the season. Schilling picked up a $333,333 check each time he didn't tip the scales too far.

6 Michael Jordan Pulls in Cash in the Minors

This last one's actually a basketball contract. Michael Jordan's abrupt departure from basketball to play minor league baseball following the Chicago Bulls' 1993 championship campaign struck most observers as odd. How could Jordan quit playing hoops and leave all of that money on the table?

As it turns out, His Airness was losing less cash than we all thought. Bulls owner Jerry Reinsdorf also owned the Chicago White Sox, the team Jordan was playing for in the minors. Even though Jordan was technically retired from basketball, Reinsdorf paid Jordan his $4 million salaries for the Bulls seasons he missed.

The first NFL pass that Brett Favre completed was caught by . . . Brett Favre (on a deflection).

BEATS PRACTICING!

FIVE DOWNRIGHT BIZARRE ATHLETE SUPERSTITIONS

1 Reserve Outfielder Has a Touching Story

Kevin Rhomberg played just 41 games in parts of three seasons with the Cleveland Indians from 1982–84. But in that short span, the outfielder managed to stake his claim as the big leagues' most superstitious player ever. Rhomberg's most peculiar superstition was that if someone touched him, he had to touch that person back.

Although this compulsion was not as much of a liability as it might have been in basketball or football, it still led to some odd situations: If Rhomberg got tagged out while running the bases, he'd wait until the defense was clearing the field at inning's end to chase down the player who'd touched him. Rhomberg also refused to make right turns while on the field because baserunners are always turning left. So if a situation forced him to make a right turn, he'd go to his left and make a full circle to get moving in the correct direction.

2 Hockey Player Gives His Stick a Swirlie

Bruce Gardiner spent five years as a forward in the NHL, most notably with the Ottawa Senators. Before each game, Gardiner would dip the blade of his stick in the locker room toilet. Gardiner's strange superstition started in his rookie reason in Ottawa in 1996. After going several games without a point, he asked veteran Tom Chorske for advice. Chorske told Gardiner he was treating his stick too well and needed to teach the wood to respect him by dunking it in the toilet.

Although Gardiner was initially skeptical, after his cold streak extended for a few more games, he took Chorske's advice. He then got hot and started scoring, and he kept on hitting the bathroom before games. Gardiner eventually backed off of dunking his stick regularly, but he'd still go back to the tactic to end a slump. As he told NHL.com in 2007, "You tape it, you dunk it, and you don't touch it. I'd do anything for a couple of goals."

Babe Ruth claimed to have only one superstition: "Whenever I hit a home run, I make certain to touch all four bases."

Reliever Breaks Out His Toothbrush

Eccentric reliever Turk Wendell pitched for four teams between 1993 and 2004 and posted some solid seasons in that span. However, he's most remembered for his vast collection of bizarre superstitions. Among Wendell's more notable quirks was his requirement that he chew four pieces of black licorice while pitching. At the end of each inning, he'd spit them out, return to the dugout, and brush his teeth, but only after taking a flying leap over the baseline.

An avid hunter, Wendell also took the mound wearing a necklace adorned with trophies from animals he had harvested, including mountain lion claws and the teeth of wild pigs and buffalo. When compared to these superstitions, Wendell's other little oddities (drawing three crosses in the dirt on the mound, always throwing the rosin bag down as hard as he could, and insisting figures in his contract end in 99 as a tribute to his jersey number) don't seem so strange.

4 Soccer Team Exorcises Its Demons

Ecuador's national soccer team knew they needed help if they were to succeed at the 2006 World Cup in Germany. Even after practicing and preparing as well as they could, they were still looking for an edge. They found it in Tzamarenda Naychapi, a mystic who London's *Guardian* called a "witch doctor-cum-shaman-cum-priest-type-fella," to help enlist the aide of supernatural spirits.

Naychapi supposedly visited each of the 12 stadiums being used in the World Cup and chased away any lingering evil spirits and worked a little magic on the pitches and goals themselves. By all accounts the spells worked; although Ecuador is not a traditional soccer powerhouse the team defeated Poland and Costa Rica in group play to advance to the Round of 16, where they lost to England 1–0.

NASCAR Drivers Get Nutty

Drivers in the top stock-car circuit have their share of superstitions, including green cars being bad luck and a hesitance to carry $50 bills. Possibly the most inexplicable, though, is their adamant refusal to deal with peanuts in their hulls. Specifically, the hulls seem to bother drivers since shelled peanuts or nuts in candy bars are perfectly kosher for the track.

No one is quite sure from where this superstition springs, but it has almost certainly been around since NASCAR's beginnings. One theory dates the tradition back to a 1937 race in Nashville in which peanut shells were sprinkled on the cars of five drivers, all of whom crashed during the race. Another possible backstory holds that one of Junior Johnson's crew was eating peanuts when an engine blew, and the blame fell on the nuts themselves. Others claim that when racing was gaining popularity in the 1930s, mechanics would often find peanut shells from the nearby grandstands in the cylinders of engines that had failed. Whatever the origin, don't take peanuts to the track with you. Any other kind of nut or legume is okay, but peanut shells will only cause misfortune.

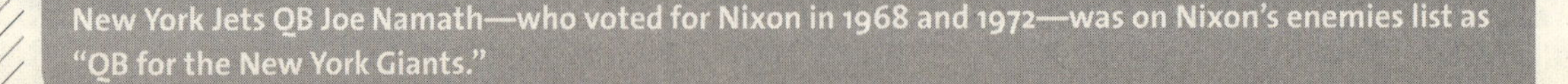

New York Jets QB Joe Namath—who voted for Nixon in 1968 and 1972—was on Nixon's enemies list as "QB for the New York Giants."

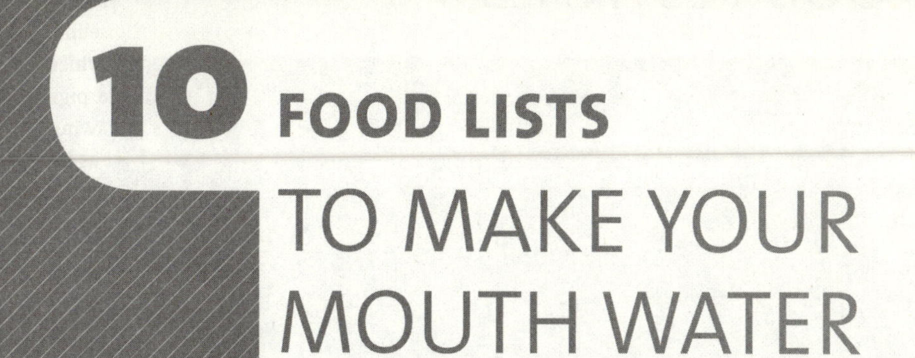

10 FOOD LISTS TO MAKE YOUR MOUTH WATER

SLOPPY JOES

THE WORLD'S EIGHT MESSIEST FOOD FESTIVALS

Italy's Orange Battle

Every year, townspeople in Ivrea, Italy, celebrate the three days before Lent by pelting one another with oranges. According to legend, the feudal lord of medieval Ivrea was so stingy that he gave his peasants only one pot of beans every six months. In protest, the villagers would throw the beans into the streets. Over the years, the beans were replaced by oranges, which grow plentifully throughout Southern Italy. The custom now known as the Orange Battle involves revelers standing on parade floats and launching the fruit at fellow participants. And it's not uncommon to see a little blood mixed in with all that orange juice. Visitors can join in, but you'll probably want to bring some goggles and a helmet.

Deep-fried scorpions on skewers are a popular street snack in China. Munching a scorpion supposedly makes your blood hotter in cold weather.

2. Night of the Radishes

When Spanish explorers brought radishes to Mexico in the 16th century, farmers near the modern-day city of Oaxaca quickly started farming the veggies. Unfortunately, nobody wanted to buy them. Not knowing what to do with all the extra produce, vendors began carving the radishes into ornate shapes and using the vegetable sculptures to lure customers to their produce stands.

Amazingly, it worked. The novelty items became so popular that farmers began leaving their radishes in the ground long after harvest season, letting them grow into bizarrely shaped behemoths. Now, December 23 is known as Noche de Rabanos (Night of the Radishes). Oaxacans celebrate it each year by gathering in the town square to display and admire elaborately detailed radishes modeled into saints, nativity scenes, and even the town itself.

3. Turkey's Greasy Wrestling Competition

The Turks sure do love their olive oil. In fact, they're so obsessed with the stuff that it plays a leading role in one of their treasured national pastimes—the Kirkpinar wrestling contest. At nearly 650 years old, the tournament is one of the world's longest continuously running sporting events. It's also one of the most popular. Each June, more than 1,000 competitors cover themselves in a slick coat of olive oil before entering the ring. All that grease makes for some comically slippery bouts, but that doesn't stop the Turks from taking this event seriously. Millions of spectators turn out for the three-day tournament, and the champion (crowned the "Big Hero") is honored as the country's preeminent sports star.

4. The West Virginia Roadkill Cook-Off

Never let it be said that West Virginians can't poke fun at themselves. The annual Roadkill Cook-Off embraces the state's hillbilly image by celebrating a 1998 law that allows people to cook any meat found on the side of the highway. The festival's motto—"You kill it, we grill it!"—sums up the menu perfectly; it's a smorgasbord of scavenger's delights, including deer fajitas, BBQ buzzard, and squirrel gravy over biscuits.

5. Greece's Clean Monday Flour War

Many parts of the world go crazy during Carnival, but in the Greek seaside town of Galaxidi, it's all about the day-after festival, known as Clean Monday. That's when locals pummel one another with bags of multicolored flour, powdering the entire town like a doughnut. The food coloring in the flour is strong enough to stain old buildings, so before they unleash more than 3,000 lbs. of the stuff in the streets, the people of Galaxidi cover much of the city in plastic.

TV foodie Alton Brown was the director of photography for R.E.M.'s 1987 video for "The One I Love."

The Mame-Maki Ritual

For centuries, the Japanese have marked the beginning of spring as a time to drive evil spirits out of their homes. The most common method for achieving this is the mame-maki ritual, during which families toss roasted soybeans around their houses and chant "bad luck out, good luck in!" At the end of the ritual, participants pick up and eat a bean for each year of their lives, assuring good fortune for the year ahead. Nowadays, children can be seen madly tossing beans onto the street, while celebrities and monks alike host parties in large temples and shower the crowds with soy.

7 The Lopburi Monkey Festival

Like many places in Thailand, the city of Lopburi is overrun with macaque monkeys. They swing freely through the streets, hitch rides on top of cars, and snatch food from the hands of unsuspecting tourists. But even though the animals are annoying, the Thais worship them. According to Hindu legend, a god named Hanuman (the Monkey King) once ruled this region. In his honor, the city celebrates once a year by feeding its 2,000-plus monkeys a huge buffet overflowing with tropical fruits, flavored rice dishes, and modern treats such as Coca-Cola.

8 Shepherd's Shemozzle

Leave it to the Kiwis to out-weird us all. Hunterville, New Zealand, is home to the Shepherd's Shemozzle, a 2-mile race in which shepherds and their dogs trek through an obstacle course that offers a different eating challenge each year. Past trails have included sheep's eyes and oil-marinated bugs, but the 2008 contest may have been the strangest of them all. Contestants had to run 50 meters while clenching raw bull testicles in their teeth. Then, before the taste was out of their mouths, they had to eat a brick of dry Weetabix cereal, followed by a raw egg and a warm can of beer.

At the chain Fatburger, you can order a "Hypocrite"—a veggie burger topped with crispy strips of bacon.

SCRAPPY MEALS

FOUR FOODS PEOPLE ACTUALLY DIE FOR

1 Iwatake

The annals of Arctic exploration are filled with accounts of frostbitten limbs and near starvation. In fact, many adventurers have reported being so hungry that they've scraped papery-crisp lichen off rocks and boiled it into passably edible food. One outdoorsman even claimed that if braised shoe leather was in a taste-test with lichen, the shoe leather would come out on top.

And yet, this very same survival food is considered a delicacy in Japan. There, iwatake (*iwa* meaning rock, and *take* meaning mushroom) is so highly sought-after that harvesters are willing to rappel down cliff faces for the precious growths. (It takes about a century for the lichen to get to a worthwhile size.)

Needless to say, this is specialty work. As if the rappelling isn't tricky enough, iwatake is best harvested in wet weather, because the moisture reduces the chance that the lichen will crumble as it's pried off with a sharp knife. In its preferred preparation, the black and slimy raw material is transformed into a delicate tempura. And while iwatake in any form doesn't taste like much, it's esteemed for its associations with longevity. As for the harvesters? Their longevity's more questionable. "Never give lodging to an iwatake hunter," goes an old Japanese adage, "for he doesn't always survive to pay rent."

2 Bird's Nest Soup

Cantilevered high off cave walls and cliffs along the seas of Southeast Asia are the nests of the white-nest swiftlet—a bird that's managed to turn an embarrassing drool problem into a useful DIY project. The nests, sturdy constructions no bigger than the palm of your hand, are made from the birds' spit. Yup, these swiftlets have specialized saliva glands powerful enough to turn their tongues into avian glue guns.

You'd think being stuck in caves high above the ground, and the fact that they're birds' nests, would protect them against humans—but no. Ever since sailors first brought the nests home for the Chinese emperor and his family in the first century CE, bird's nest soup has been a favorite among the country's elite. Never mind that it's virtually tasteless; the dish is revered for health reasons.

Of course, acquiring the main ingredient is less healthy. Nest harvesters must stand on rickety bamboo scaffolding hundreds of feet off the ground in pitch darkness. They must also endure unbelievable heat and humidity as they try to avoid all the insects, birds, and bats that live in the caves. In addition, the extraordinary value of the nests means the zones are

Edam cheese balls were occasionally fired off ships in the Caribbean Sea as cannonballs in the 17th and 18th centuries.

patrolled by machine-gun toting guards. Harvesting rights are multiyear, multimillion-dollar deals arranged with national governments, and poaching is ruthlessly prohibited. Unarmed fishermen have been shot dead after accidentally beaching in swiftlet territory, and local tour group operators pay exorbitant fees to avoid rifle-assisted leaks springing in their kayaks. It all underscores the fact that being a nest harvester is less of a career choice and more of a life sentence—especially considering that the skill is almost exclusively passed on from father to son.

Snapping Turtle

Turtle soup was a staple of 19th-century gourmets, usually ladled out of huge tureens for the first course. And no wonder; turtle meat is tasty, fibrous, and chewy—kind of like barbecued pork. But getting the meat in the quantities Grover Cleveland and his ilk demanded meant getting the biggest turtles around, and in most of the United States, that meant going after snapping turtles. The traditional means of capturing the giant creatures (which grow up to 180 lbs.) is called noodling, which involves brave souls trawling along the banks of rivers, lakes, and ponds, and occasionally wading neck-deep to stick a boot into the turtles' lairs. If a noodler hits shell, next in are the hands, which try to haul the critter out while avoiding its famously strong jaws.

On-the-job accidents come with the territory. According to outdoor expert Keith Sutton, author of *Hunting Arkansas*, "noodlers are nicknamed 'nubbins' as the result of unfortunate encounters with snappers." Amazingly, the job isn't over once the turtle is captured, either. Turns out, killing the animal is another exercise in raw nerve. We'll spare you the details, except to say that it's ill-advised to handle the animal's head until at least a day after its execution. Even decapitated, the snapping turtle has a long memory.

Gooseneck Barnacle

You've probably never seen gooseneck barnacles on a menu in the States, but it's only a matter of time. Besides being a popular Christmastime appetizer in Spain and Portugal (where it's known as *percebes*), it's gaining ground in America and being harvested off the coast of the Pacific Northwest.

Harvesting this rock-dwelling crustacean is no simple matter, though. Barnacle fishers typically tie themselves to the rocks in a surge zone along the ocean and pry the creatures off between waves. To do this, they have to use a crowbar to break the animals' self-adhesive, which is so resistant to tampering that scientists were long mystified by its chemical makeup. In other words, removing a barnacle takes lots of traction, which, given the waves, can be tricky. A poorly maintained tether, or a harvester too impatient to tie in, can easily end with a call to the Coast Guard. Of those brave enough to harvest gooseneck barnacles, one Coast Guard official said, "The best we can do is retrieve the bodies."

The first cow to ride in an airplane was Elm Farm Ollie in 1930. The milk she gave in-flight parachuted down over St. Louis.

EIGHT FOODS THAT (THANKFULLY) FLOPPED

Coffee-Flavored Jell-O (Celery, Too!)

In 1918, the makers of Jell-O introduced a new flavor: coffee. Its release was ostensibly based on the logic that, since lots of people like to drink coffee with dessert, they'd be game for combining the two after-dinner treats. Not the case. The company soon realized if anyone wants dessert coffee, they're going to have a cup of it. In fact, if anyone wants coffee at all, they're going to have a cup of it. Not surprisingly, this realization came about the time they yanked the product off the shelves. Coffee wasn't Jell-O's only misstep: Cola-flavored Jell-O was sold for about a year starting in 1942, and for a brief while, the clear, wiggly dessert was sold in celery and chocolate flavors, too.

Horsemeat was on the menu of the Harvard Faculty Club until 1985. The meat came from nearby Suffolk Downs racetrack.

Reddi-Bacon

Any company smart enough to bless mankind with sprayable whipped cream—the sort that promotes direct-to-mouth feeding—has got to know a thing or two about immediate gratification. But sadly, the makers of Reddi-wip were unable to meld their keen understanding of human laziness with one of processed meat. They figured, if you're cooking breakfast in the morning and you've got a hankering for bacon, why dirty up a pan you'll only have to clean later? The solution: foil-wrapped Reddi-Bacon you could pop into your toaster for piping-hot pork in minutes.

While it seemed perfect for the busy 1970s household, the absorbent pad designed to soak up the dripping grease tended to leak, creating not only a fire hazard, but also a messy (if not totally ruined) toaster. Ultimately, the product lasted about as long as it took to cook; the company scrapped it before it went to market nationwide.

Cereal Mates

Sometimes, new products fail because they're simply bad ideas (ahem, New Coke). Other times, it's because they're just impossible to market. Such was the case for Cereal Mates. Beating the dead horse of uber-convenient breakfast foods, Kellogg's introduced Cereal Mates in 1997. The idea was simple: a small box of cereal, a container of specially packaged milk (no refrigeration required!), and a plastic spoon. It was the perfect A.M. answer for the person on the go . . . who enjoys warm milk on cereal. Trying to patch up one mistake with another, Kellogg's then moved the product to the dairy section, where no sane person looks for cereal. On top of all that was the price. At about $1.50 for only four ounces of the stuff, Cereal Mates was deemed too expensive for most consumers. After two years, Kellogg's pulled it from the shelves.

Gerber Singles

At some point in time, almost every adult has tasted baby food and discovered that the stuff isn't half bad. But that doesn't mean people want to make a meal out of it. For some reason, Gerber had to learn that lesson the hard way. In 1974, the company released Gerber Singles, small servings of food meant for single adults, packaged in jars that were almost identical to those used for baby food. It didn't take long for Gerber execs to figure out that most consumers, unless they were less than a year old, couldn't get used to eating a pureed meal out of a jar—particularly one depressingly labeled "Singles." Baby food for grown-ups was pulled from the marketplace shortly after its birth.

Gatorade's inventor later created an alcoholic variation, Hop 'n Gator—essentially, lemon lime Gatorade mixed with beer.

5 "I Hate Peas!"

For as long as children have been shoving brussels sprouts under mashed potatoes and slipping green beans to the dog, parents have been hunting desperately for a way to end the vegetable discrimination. Finally, in the 1970s, American Kitchen Foods, Inc. came to the rescue (or at least tried) with the release of "I Hate Peas!" Since kids love French fries so much, the company decided that disguising peas in a fry-shaped form was a surefire way to trick tots into getting their vitamins. Not a chance. Children all over America saw through the ruse. After all, a pea is a pea is a pea, and the name of the product was more than apropos, no matter what it looked like. There were other thinly disguised vegetables in the company's "I Hate" line, but kids hated those, too.

6 Heublein's Wine & Dine

In the mid-1970s, Heublein introduced Wine & Dine, an upscale, easy-to-make dinner that included a small bottle of vino. How refined. How decadent. How confusing. Consumers knew Heublein for their liquor and wines, so how were they supposed to know the wine included in Wine & Dine was an ingredient for the pasta sauce? Hasty consumers who didn't read the directions closely ended up pouring the contents of the bottle into a nice glass and getting a less-than-pleasant mouthful of salted wine.

7 Funky Fries

In 2002, hoping to follow the success of Heinz's new "kiddie" ketchup versions (in green and purple), Ore-Ida introduced Funky Fries: chocolate-flavored, cinnamon-flavored, and blue-colored French fries. An awful lot of money was sunk into the product, but after a year of marketing, consumers still found the idea funky—in the bad way. Funky Fries were pulled off the shelves in 2003, and images of blue fries with green ketchup were once again relegated to the world of Warhol-esque pop art.

Pepsi A.M.

Creating a super-caffeinated soda worked well for the makers of Red Bull, but not for the folks at Pepsi. With 25 percent more caffeine than a cup of Joe, PepsiCo introduced the cola-flavored product in 1989, only to discover that most people just couldn't bring themselves to drink soda with their cornflakes. For those who wanted a Pepsi in the morning, regular Pepsi did just fine, thankyouverymuch. Pepsi A.M., like the coffee-flavored Pepsi Kona before it, was scrapped after just a few months.

In 1907, an ad campaign for Kellogg's Corn Flakes offered a free box of cereal to any woman who would wink at her grocer.

TASTEMAKERS

THE BIRTHPLACES OF SEVEN GREAT AMERICAN FOODS

1. **The Hamburger**
Louis' Lunch, New Haven, Conn.

There are competing claims for the coveted "Inventor of the Hamburger" title, but according to Louis' Lunch (and the Library of Congress, for that matter), this small New Haven restaurant takes the prize. The story goes something like this: One day in 1900, a rushed businessman asked owner Louis Lassen for something quick that he could eat on the run. Lassen cooked up a beef patty, put it between some bread, and sent the man on his way.

Pretty modest beginnings for arguably the most popular sandwich of all-time, huh? If you visit Louis' today, you'll find that not much has changed. The Lassen family still owns and operates the restaurant, the burgers are still cooked in ancient gas stoves, and, just like then, there is absolutely no ketchup allowed.

2. **The Root Beer Float**
Myers Avenue Red Soda Co., Cripple Creek, Colo.

If you thought what happened up on Cripple Creek only happened in song, you're sorely mistaken. In August 1893, a failed gold-miner-turned-soda-company-owner named Frank J. Wisner was drinking a bottle of his Myers Avenue Red root beer while looking up at Cow Mountain. Just then, a full moon illuminated the snowcap on the otherwise black mountain, and Wisner had a brilliant idea—float a scoop of vanilla ice cream in a glass of his root beer. The new drink was christened the "black cow" and became an instant classic. Today, of course, most of us call it a root beer float.

3. **The Corn Dog**
Cozy Dog Drive In, Springfield, Ill.

In 1946, Ed Waldmire, Jr., revolutionized the stick-meat world when he debuted the Cozy Dog—the first corn dog on a stick. At first, he wanted to call his creation the "Crusty Cur," but his wife convinced him to change the name to "Cozy Dog." She felt people wouldn't want to eat something described as "crusty." Good call, Mrs. Waldmire. Shortly after the Cozy Dog's inception, the Cozy Dog Drive In opened alongside old Route 66 and has been serving up corn dogs ever since.

The average American child eats about 1,500 peanut butter and jelly sandwiches before finishing high school.

4 The Philly Cheesesteak
Pat's King of Steaks, Philadelphia, Pa.

Philadelphia is known for many things (Ben Franklin, the Liberty Bell, and Rocky, for starters), but fine dining is not really its forte. That's OK, though, because Philly is the home of Pat's King of Steaks, and Pat's King of Steaks is where the Philly cheesesteak was born. One day back in 1932, hot-dog-stand owners Pasquale (Pat) and Harry Olivieri decided to change things up and make a steak sandwich with onions. A cabdriver who ate at Pat's daily insisted on trying the new sandwich, and with the first bite declared, "Hey, forget 'bout those hot dogs, you should sell these!"

The code name used for the microwave while it was still in testing was "Speedy Weenie."

Cabdrivers know fast food about as well as anyone, so the brothers did just what the cabbie suggested. In no time, the modest stand turned into the Pat's that exists today. Controversy remains, however, over who's responsible for putting the cheese in cheesesteak. Pat's claims it was the first to do so (in 1951), but across-the-street rival Joe Vento of Geno's Steaks (opened 1966) insists he added the finishing touches.

The Onion Ring
Pig Stand, Dallas, Texas

According to most sources, the onion ring was invented when a careless cook at a Pig Stand location in Dallas accidentally dropped an onion slice in some batter, then pulled it out and tossed it in the fryer for lack of a better destination. Now, you'd think inventing the onion ring would be enough for one restaurant chain, but not Pig Stand. The company also lays claim to opening America's first drive-in, inventing Texas toast, and being one of the first restaurants to advertise using neon signs. Not bad for a little outfit from Texas.

6 The Pizzeria
Lombardi's, New York, N.Y.

Pizza has existed in one form or another for a long time, but America got her first true pizzeria when Gennaro Lombardi opened up a small grocery store in NYC's Little Italy. An employee named Anthony "Totonno" Pero started selling pizzas out of the back, and in no time, Lombardi's was concentrating on its burgeoning pizza business instead of plain old groceries.

In 1905, the establishment was licensed as a pizzeria, and it's stayed that way ever since. Well, almost. The original restaurant closed in 1984 but reopened down the street 10 years later. On its 100th anniversary in 2005, Lombardi's decided to offer its pizza for the same price it'd been sold for in 1905—5 cents a pie. Needless to say, the line wrapped around the block.

7 The Fried Twinkie
The ChipShop, Brooklyn, N.Y.

Sometimes what counts isn't being the inventor, it's being the innovator. Take the fried Twinkie, for example. The Twinkie—in all its indestructible glory—has been around for ages, but when ChipShop owner Christopher Sell had the brilliant idea to freeze the snack, dip it in batter, and deep-fry it, the Twinkie took gluttony to new heights. Even the *New York Times* raved about how "something magical" happens when you taste the deep-fried Twinkie's "luscious vanilla flavor." Sell, who was trained in classical French cuisine, didn't start with the Twinkie, though. In his native England, he fried up everything from M&M's to Mars bars.

Waste from the Ben & Jerry's factory is given to farmers to feed their hogs. The hogs love Cherry Garcia but dislike Mint Oreo.

IS THERE A MR. TWINKIE?

SIX TASTY FOODS NAMED AFTER PEOPLE

1 Salisbury Steak

James Salisbury was a 19th-century American doctor with a rather kooky set of beliefs. According to Salisbury, fruits, vegetables, and starches were the absolute worst thing a person could eat, as they would produce toxins as our bodies digested them. The solution? A diet heavy on lean meats. To help his cause, Salisbury invented the Salisbury steak, which he recommended patients eat three times a day and wash down with a glass of hot water to aid digestion. Apparently the only people paying attention to the doctor's orders were elementary school lunch ladies.

2 Graham Crackers

Sylvester Graham would not have gotten along very well with James Salisbury. Graham, a 19th-century diet proponent, felt that people should ingest mostly fruits, vegetables, and whole grains while avoiding meats and any sort of spice. The upside of all of this bland food sounds a bit curious to the modern reader: Graham thought his diet would keep his patients from having impure thoughts. Cleaner thoughts would lead to less masturbation, which would in turn help stave off blindness, pulmonary problems, and a whole host of other potential pitfalls that stemmed from moral corruption. Graham invented the cracker that bears his name as one of the staples of this anti-self-abuse diet.

3 Nachos

In 1943 Ignacio Anaya—better known by his nickname "Nacho"—was working at the Victory Club in Piedras Negras, Mexico, just over the border from Eagle Pass, Texas. As the story goes, there were a lot of American servicemen stationed at Fort Duncan near Eagle Pass, and one evening a large group of soldiers' wives came into Nacho's restaurant as he was closing down.

Nacho didn't want to turn the women away with empty stomachs, but he was too low on provisions to make a full dinner. So he improvised. Nacho Anaya supposedly cut up a bunch of tortillas, sprinkled them with cheddar and jalapeños and popped them in the oven. The women were so delighted with the *nachos especiales* that the snack quickly spread throughout Texas.

Ben & Jerry originally wanted to get into the bagel business, but the equipment was too expensive. So they settled on ice cream.

Bananas Foster

In 1951, Richard Foster had a tough job. He was the chairman of a New Orleans crime commission that was trying to clean up the French Quarter, and he also ran his own business, the Foster Awning Company. When Foster was hungry, he would often head into his friend Owen Brennan's restaurant, Brennan's, and happily wolf down whatever chef Paul Blangé was making. When Chef Blangé invented a new dessert of flaming bananas, he named it after his owner's buddy and frequent customer.

Fettuccine Alfredo

The Italian favorite has been around for centuries, but it supposedly took on its current form around 1914 when Alfredo di Lelio upped the amount of butter in the recipe in an attempt to find something his pregnant wife would enjoy eating. Di Lelio realized that his buttery cheese sauce was extraordinarily tasty, so he started serving it to tourists at his Rome restaurant and named the dish after himself.

6 Margherita Pizza

This deliciously simple pizza is named after Margherita of Savoy, who was Queen consort of Italy from 1878 until 1900 during the reign of her husband, King Umberto I. In 1889, Umberto and Margherita took a vacation to Naples and visited renowned pizza chef Raffaele Esposito, who cooked the royal couple three special pizzas. Margherita particularly enjoyed one that had used mozzarella, tomato, and basil to mimic the colors of the Italian flag, so Esposito named the dish in her honor.

Three Real People in Your Pantry

1. Duncan Hines was a notable restaurant and hotel critic in the mid-20th century before licensing his name to snack foods.
2. Little Debbie was the then-4-year-old granddaughter of the snack cake company's founders. That's her picture on the box!
3. Mrs. Fields has been cranking out delicious cookies since young mother Debbi Fields opened her first store in Palo Alto, Calif. in 1977.

. . . and Two That Aren't

1. Aunt Jemima pancake mix originally had the catch name "Self-Rising Pancake Flour," but the name changed in 1899 thanks to a popular minstrel song "Old Aunt Jemima."
2. Betty Crocker is a marketing invention. The Washburn-Crosby Company selected "Betty" because it sounded cheery and "Crocker" to honor an executive who had just retired.

Fredric Baur invented the Pringles can. When he died in 2008, his ashes were buried in one.

WHERE ARE THESE THOUSAND ISLANDS?

THE ORIGINS OF FIVE CONDIMENTS

1 Thousand Island Dressing

Is the delicious dressing that gives a Reuben its tanginess named after an actual chain of islands? You bet it is. The Thousand Islands are an archipelago that sits in the Saint Lawrence River on the U.S.–Canada border, and there are actually well over a thousand of them, some of which are so small that they contain nothing more than a single home.

So why is the dressing named after an archipelago? No one's quite sure. Some people claim that early film star and vaudevillian May Irwin, who summered on the Thousand Islands, named it, while others contend that George Boldt, the famed proprietor of the Waldorf-Astoria, gave the dressing its name because of his own summer place in the region. No matter who named it, it's tough to beat on a sandwich.

2 Ranch Dressing

Yep, the beloved dressing and dipping sauce actually got its start on a real ranch. When Steve and Gayle Henson opened a dude ranch in California in 1954, they had an ace up their sleeves: a delicious dressing that Steve had concocted while the couple was living in Alaska.

The couple did a nice business at their Hidden Valley Ranch, but guests were always flipping out over just how tasty Steve's dressing was. Eventually, the Hensons started selling dried packets of the stuff, and the popularity grew so quickly that they had to hire a 12-man crew just to help mix up each batch. Steve's culinary creativity turned out to be lucrative; in 1972 Clorox forked over $8 million for the recipe.

3 A1 Steak Sauce

According to the brand's website, A1 has been around for quite a while. Henderson William Brand worked as the personal chef for King George IV from 1824 to 1831, and at some point during this employment he mixed up a new sauce for the king to use on his beef. George IV allegedly took one bite of Brand's creation and declared that it was "A1." Brand then left the king's employ in order to go peddle his new sauce.

Procter & Gamble chose the "Pringles" name from a Cincinnati phone book. They liked the name of Pringle Drive.

Worcestershire Sauce

Worcestershire sauce was invented accidentally in England by Brits trying to ape what they thought was authentic Indian food. In this case, the demanding diner was one Lord Marcus Sandy, a former colonial governor of Bengal. Having grown attached to a particular flavor of Indian sauce, he recruited two drugstore owners, John Lea and William Perrins, in hopes that they could re-create it based on his descriptions. Lea and Perrins thought they'd make a profit by selling the leftovers in their store, but unfortunately, the sauce they created had a powerful stench. They stashed the stinky sauce in the basement and forgot about it for two years while it aged into something that tasted and smelled much better. (We suspect that in a similar manner, we are harboring the next big culinary phenomenon in the back of our fridge.)

5 Heinz 57

Legend has it that Heinz 57 takes its name from the H. J. Heinz company formerly marketing 57 products at once, but the number's not quite right. Heinz's website tells the story that Henry John Heinz was riding a train when he saw a billboard advertising 21 varieties of shoes. He so liked the idea he wanted to try it with his own condiment company. Thus, he started touting Heinz's 57 varieties.

There was only one catch: Heinz marketed well over 60 products at the time. So where did the 57 come from? Heinz thought the number was lucky. Five was Heinz's lucky number, and seven was his wife's. He mashed the charmed digits together, got 57, and never looked back.

Cashews are one of a few nuts never sold in their shells. Their shells contain a toxic liquid that causes skin rashes.

10 ESSENTIAL TALKING POINTS

FOR THE BANANA ENTHUSIAST

1. In recorded history, bananas date to around 600 BCE, when they were mentioned in Buddhist texts.

2. Bananas came to the Americas in the 15th century but weren't regularly imported to the United States until after their showcase at the 1876 Philadelphia Centennial Exhibition.

3. The average American eats 30 pounds of bananas a year.

Twinkies originally had banana-flavored filling but switched to vanilla when WWII brought the banana trade to a halt.

As a kid, Lucille Ball was fired from a drugstore for forgetting to put bananas in a banana split.

5. Banana peels have been a staple of slapstick slip-and-fall gags for decades, but in the 1960s, many people tried using them for a different purpose—to get high. Rumors that smoking dried banana peels caused hallucinogenic effects were likely started in part by singer Country Joe McDonald, who mistakenly attributed an acid trip to a banana-peel joint he'd tried. Regardless, the trippy allegations touched off a banana run on fruit stands across the country, until an FDA investigation found no evidence to support the claim.

6. Because they're a major source of potassium, vitamin C, and other nutrients, bananas are believed to aid in the treatment of morning sickness and hangovers.

7. As any banana aficionado knows, a bunch of bananas is called a hand; individual bananas are called fingers.

8. That luscious yellow fruit isn't the only edible part of the banana plant. In fact, the banana flower often finds its way into soups and curries in Southeast Asia, while the succulent core of the plant's trunk is sometimes used in Burmese and Bengali cuisine. Even the leaves can be handy in the kitchen—they're often used to wrap food while it's being steamed.

9. Like money, bananas don't grow on trees. They grow on plants. Unlike money, they're members of the same family as Manila hemp.

Woody Allen reportedly only named his 1971 political satire film *Bananas* because "there are no bananas in it."

Why is bubble gum pink? When it was invented, pink was the only food dye on hand.

GET US TOM COLLINS!

THE STORIES BEHIND SEVEN COCKTAILS

1 The Manhattan

Scores of people claim to have invented the venerable Manhattan, a blend of whiskey, sweet vermouth, and bitters. It may date back to the New York bar scene of the 1860s, but there are also some more intriguing tales about its origins. According to one of these legends, Jennie Churchill threw a party at the Manhattan Club in 1874 to celebrate Samuel J. Tilden's victory in New York's gubernatorial election. An enterprising bartender created a new cocktail for the event, which he dubbed the Manhattan in the club's honor.

Both Churchill and Tilden would go on to bigger things once they sobered up. Churchill soon gave birth to a son, Winston, and Tilden made a presidential run in 1876. (Although Tilden won the popular vote, he lost out to his Republican opponent, Rutherford B. Hayes. At least the cocktail saved him from obscurity.)

2 The Tom Collins

This refreshing summer drink owes its name to a 19th century hoax. In 1874, hundreds of New Yorkers heard some bad news while they were out on the town: a certain Tom Collins had been besmirching their good names. Although these people didn't know Mr. Collins, they were outraged that he would slander them, and they often set out to find the rascal.

Of course, the root of the hoax was that there wasn't really a Tom Collins, but that didn't keep aggrieved parties from searching him out. To deepen the joke, bartenders started making the citrus cocktail that now bears the name, so when searchers asked for Tom Collins, they could instead find a thirst-quenching long drink.

The Negroni

Count Camillo Negroni gets credit for creating this aperitif around 1919. As the story goes, Negroni really loved to throw back an Americano (Campari, sweet vermouth, and club soda), but he wanted a little extra zing in his glass. He asked a bartender to replace the club soda with gin to give the mixture some added kick, and the Negroni was born.

Until coffee gained popularity, beer was the breakfast beverage of choice in most urban areas of the United States.

Long Island Iced Tea

It might not actually contain tea, but at least the Long Island part of the name is accurate. This spring break favorite is fairly young as cocktails go; it's only been around since 1976. Rosebud Butt, a bartender at the Oak Beach Inn in Hampton Bays, invented the drink, so if you ever need to find a patron saint of terrible hangovers and nights spent falling off barstools, Rosebud may be your man.

The Kir

This aperitif of crème de cassis and white wine has long been a favorite in France, but it didn't get its name until after World War II. Felix Kir, the mayor of Dijon from 1945 to 1968, was a huge fan of the cocktail, and whenever he entertained visiting dignitaries, he'd invariably serve them the drink. Kir did such a good job pushing the mixture onto his visitors that it became inextricably linked with his personality, and that's why the cocktail bears his name today.

The Bellini

This delightful wine cocktail, a blend of white peach puree and Prosecco, has a well-established origin. Giuseppe Cipriani, founder of Venice's beloved Harry's Bar, started mixing up the fruity tipples sometime between 1934 and 1948. The pink drink reminded him of the color of a saint's toga in a painting by Italian Renaissance artist Giovanni Bellini, so Cipriani named his concoction in honor of the painter.

The Martini

The debate over the true origin of the martini can be just as contentious as the one over the proper ratio of gin to vermouth. Some claim that the martini is simply a dryer version of an older cocktail called the Martinez; Martinez, California, the birthplace of this cocktail, thus stakes its claim to the title of Birthplace of the Martini. Others postulate that the drink's name simply comes from Martini & Rossi, an Italian company that's been exporting its vermouths to the United States since the 19th century. Still others claim that the drink was created by and named for Martini di Arma di Taggia, the bartender at New York's Knickerbocker Hotel, although there's evidence that the cocktail may have been invented well before he started mixing drinks.

How many licks does it take to get to the center of a Tootsie Pop? On average, a licking machine made at Purdue University needed 364.

UNDER THE TABLE

THREE DRINKING STORIES THAT PUT YOURS TO SHAME

1 Admiral Edward Russell's 17th-Century Throwdown

Think you can drink like a sailor? Maybe you should take a moment to reflect on what that truly means.

The record for history's largest cocktail belongs to British Lord Admiral Edward Russell. In 1694, he threw an officer's party that employed a garden's fountain as the punch bowl.

The concoction? A mixture that included 250 gallons of brandy, 125 gallons of Malaga wine, 1,400 pounds of sugar, 2,500 lemons, 20 gallons of lime juice, and 5 pounds of nutmeg.

A series of bartenders actually paddled around in a small wooden canoe, filling up guests' cups. Not only that, but they had to work in 15-minute shifts to avoid being overcome by the fumes and falling overboard.

The party continued nonstop for a full week, pausing only briefly during rainstorms to erect a silk canopy over the punch to keep it from getting watered down. In fact, the festivities didn't end until the fountain had been drunk completely dry.

2 The London Brew-nami of 1814

The Industrial Revolution wasn't all steam engines and textile mills. Beer production increased exponentially, as well. Fortunately, the good people of England were up to the challenge and drained kegs as fast as they were made. Brewery owners became known as "beer barons," and they spent their newfound wealth in an age-old manner—by trying to party more than the next guy.

Case in point: In 1814, Meux's Horse Shoe Brewery in London constructed a brewing vat that was 22 feet tall and 60 feet in diameter, with an interior big enough to seat 200 for dinner—which is exactly how its completion was celebrated. (Why 200? Because a rival had built a vat that seated 100, of course.)

After the dinner, the vat was filled to its 4,000-barrel capacity. Pretty impressive, given the grand scale of the project, but pretty unfortunate given that they overlooked a faulty supporting hoop. Yup, the vat ruptured, causing other vats to break, and the resulting commotion was heard up to 5 miles away.

A wall of 1.3 million gallons of dark beer washed down the street, caving in two buildings and killing nine people by means of "drowning, injury, poisoning by the porter fumes, or drunkenness."

There really was a Captain Morgan. He was a Welsh pirate who later became the lieutenant governor of Jamaica.

The story gets even more unbelievable, though. Rescue attempts were blocked and delayed by the thousands who flocked to the area to drink directly off the road. And when survivors were finally brought to the hospital, the other patients became convinced from the smell that the hospital was serving beer to every ward except theirs. A riot broke out, and even more people were left injured.

Sadly, this incident was not deemed tragic enough at the time to merit an annual memorial service and/or reenactment.

New York State of Mind: The Dutch Ingratiate Themselves to the Natives

In 1609, the Dutch sent English explorer Henry Hudson westward for a third attempt at finding the fabled Northeast Passage. A near mutiny forced him southward, and upon reaching land, he encountered members of the Delaware Indian tribe.

To foster good relations, Hudson shared his brandy with the tribal chief, who soon passed out. But upon waking up the next day, he asked Hudson to pour some more for the rest of his tribe. From then on, the Indians referred to the island as Manahachtanienk — literally, "The High Island."

And not "high" as in "tall"; high as in "the place where we got blotto." Most people would agree that Manhattan has stayed true to the spirit of its name ever since.

After Chicago's Great Fire of 1871, brewer J. Schlitz made a unique donation to the recovery effort: hundreds of barrels of beer for thirsty Chicagoans.

10 LISTS
THAT MEAN BUSINESS

SIX JOBS YOU DIDN'T HEAR ABOUT ON CAREER DAY

In preschool, there are only a handful of sensible options for the career-minded four-year-old: doctor, plumber, firefighter, and astronaut. Clearly, had we heard about "knocking up" or any of these other fine options, we would have eaten more paste and focused less on our permanent records.

Filibuster

Long before the term "filibuster" came to be associated with elected officials, it was associated with violence and trickery. (Wait a second . . .) In the 1600s, pirates known to the Dutch as *vrijbuiters* pillaged the West Indies, and eventually, the word was assimilated into the English language as "filibusters."

Between 1850 and 1860, the name was used to refer to the American mercenaries who attempted to revolutionize Central America and the Spanish West Indies. The most famous of these filibusters was William Walker, a U.S. citizen who succeeded in gaining control of Nicaragua in 1856 by overthrowing the nation's administration. Walker became president of Nicaragua, but only until May 1, 1857, when a coalition of Central American states ousted him. Because filibusters of previous centuries strove to interfere with foreign regimes, the term evolved to refer to anyone who attempted to obstruct the government, as our legislators occasionally see fit to do when a particularly troublesome bill comes before them.

2 Lungs

Perhaps the cruelest case of naming irony in history, anyone employed to fan the fire in an alchemist's workshop was known as a "lungs." And because most alchemists were constantly trying to make gold out of lead and other such base metals, you can only imagine what kinds of dangerous materials were floating about in the labs. As a result, the actual lungs on a lungs gave out relatively quickly, leading to a profession with widespread early retirement.

There are 293 different ways to make change for a dollar.

Sin-Eater

No matter how much you loved Grandma and Grandpa, you can probably admit your forebears weren't perfect. So, if you ever had a loved one that passed on before his or her last chance at absolution, it makes sense that you might want to call in reinforcements. Fortunately for the fretful and grieving of yore, there was the town sin-eater. For a small fee, the sin-eater would gladly scarf down a meal (usually bread and ale) that had been placed on the deceased's chest. By letting the food lie atop the dearly departed for a while, it was believed the vittles would absorb the last transgressions. And, once the food was gobbled up by the sin-eater, Gramps could get into heaven without any major roadblocks.

Knocker-Up

In British towns of yore, particularly those with a mine or mill as the center of commercial activity, knocker-ups were responsible for going from house to house to wake workers in the mornings. The title, not surprisingly, came from the sound they made rapping on windows. As for the evolution of the term "knocking," it also denoted a collision of sorts, and in the 17th century, it was used in reference to childbirth. Even poet John Keats wrote of "knocking out" children in some of his odes. It wasn't until the 19th century, however, that Americans began using the phrase as slang for getting a woman pregnant.

5

Gong Farmer

Not unlike *The Gong Show*, a gong farmer was far from being the cream of the crop—and even that might be the understatement of the year. In Tudor England, a gong farmer's job was to empty the town toilets. But the job did have its perks. Typically, a gong farmer would "mine" the waste for any items of value that might be found among the city's excrement—a penny here, a button there—before it was used as manure or thrown into the river. For a while, it was falsely believed that gong farmers were immune to the plague, but you can't help wonder if that was more of a pity belief, like the whole idea that being hit by bird droppings is good luck.

6

Badger

Odd as it may sound, badgers were part of the rat race in prior centuries, serving as intermediaries between the producers of goods and the consumer. Most often, they traded in corn and other foodstuffs, buying from farmers and reselling the goods at markets in town. And if you think the salespeople at Macy's are tough, some historians think badgers were so persistent in pushing their products that the term came to be associated with an often annoying and forceful adamance—i.e., "badgering" anyone in sight to buy from you instead of another vendor.

Since 1982 every man, woman, and child who has lived in Alaska for a full 12 months gets an annual check. In 2010 each person pulled in $1,281.

HIGH INITIAL RETURNS

10 ABBREVIATED COMPANY NAMES EXPLAINED

Dozens of companies use initials in their names, but how well do you know what the letters mean?

CVS

Sorry, drugstore fans, there aren't three fatcat pharmacists with these initials running around out there. When the pharmacy chain was founded in Lowell, Mass., in 1963, it was known as "Consumer Value Stores." Over time the name became abbreviated to simply CVS.

2

Kmart

Longtime five-and-dime mogul Sebastian S. Kresge opened his first larger store in Garden City, Michigan, in 1962. The store was named Kmart after him. (Kresge had earned the right to have a store named for him; he opened up his new venture at the tender age of 94.)

The mermaid in Starbucks' logo originally had bare breasts. When Howard Schultz bought the chain in 1987 he changed the logo so her long, flowing hair covered the naughty bits.

3 IKEA

The Swedish furniture giant takes its name from founder Ingvar Kamprad's initials conjoined with the first initial of the farm where Kamprad grew up, Elmtaryd, and the parish he calls home, Agunnaryd.

4 3M

The conglomerate behind Post-it Notes gets its name from its roots as a company that mined stone to make grinding wheels. Since it was located in Two Harbors, Minn., the company was known as Minnesota Mining and Manufacturing, which was later shortened to 3M.

5 H&M

The beloved clothing store began in Sweden in 1947. Founder Erling Persson was selling only women's duds, so he called the store Hennes—Swedish for "hers." Twenty-one years later, he bought up a hunting supplier called Mauritz Widforss. After the acquisition, Persson branched out into men's clothing and began calling the store Hennes and Mauritz, which eventually became shortened to H&M.

6 GEICO

The adorable gecko's employer is more formally known as the Government Employees Insurance Company. Although GEICO has always been a private, standalone company, its name reflects its original purpose: Married couple Leo and Lillian Goodwin founded the company in 1936 to sell insurance directly to employees of the federal government.

7 ING Group

The banking giant's name is an abbreviation of Internationale Nederlanden Groep, or "International Netherlands Group," a nod to the company's Dutch origins and headquarters. The company's heavy use of the color orange in its buildings and promotion is also a shout-out to the Netherlands; orange is the color of the Dutch royal family dating all the way back to William of Orange.

It's estimated that 10% of living Europeans were conceived on an IKEA-produced bed.

8 H&R Block

Brothers Henry and Richard Bloch founded the tax preparation firm in Kansas City in 1955. Their only problem was their last name. The brothers worried that people would mispronounce their surname as "blotch," hardly a term you want associated with your tax return. They decided to sidestep this problem by spelling the company's name "Block" instead, so that nobody would miss the solid hard "k" sound.

9 YKK

The initials you see on darn near every zipper you own stand for the Japanese name Yoshida Kogyo Kabushikikaisha, which translates roughly into "Yoshida Company Limited." The company is named after Tadao Yoshida, who started the zipper concern in Tokyo in 1934.

10 P. F. Chang's

If you go looking for Mr. P. F. Chang, you'll be in for a long search. The Asian dining chain's name is actually a composite of the founding restaurateur Paul Fleming's initials and a simplification of founding chef Philip Chiang's last name.

In 1998, a Georgia teen was suspended for one day for wearing a Pepsi T-shirt at his school's "Coke in Education Day."

WALLETS AND GRIMACE

FIVE WILDLY SUCCESSFUL PEOPLE WHO SURVIVED BANKRUPTCY

Not being able to pay your creditors is never a good situation. But it may not be the financial kiss of death you think. Here are a few famous people who clawed their way out of the poorhouse.

Abraham Lincoln

His face may now appear on the penny, but at one time, Lincoln didn't have a single cent to spare. Lincoln tried many occupations as a young man, including buying a general store in New Salem, Ill., in 1832. While he may have been terrific at splitting rails, winning debates, and wearing stovepipe hats, Honest Abe wasn't much of a shopkeeper. Lincoln and his partner started buying out other stores' inventories on credit, but their own sales were dismal.

During the Depression, La-Z-Boy sold chairs on a barter system. Coal, wheat, or even chickens could get you a recliner.

As the store's debts mounted, Lincoln sold his share, but when his partner died, the future president became liable for $1,000 in back payments. Lincoln didn't have modern bankruptcy laws to protect him, so when his creditors took him to court, he lost his two remaining assets: a horse and some surveying gear. That wasn't enough to foot his bill, though, and Lincoln continued paying off his debts until well into the 1840s.

Other presidents: Lincoln's not alone in the annals of bankrupt commanders-in-chief. Ulysses S. Grant went bankrupt after leaving office when a partner in an investment-banking venture swindled him. Thomas Jefferson faced bankruptcy several times, including after leaving office, possibly because he threw around a lot of cash on food and wine. William McKinley went bankrupt while serving as Ohio's governor in 1893; he was $130,000 in the red before eventually straightening out with the help of friends. He won the White House just three years later.

Henry Ford

Ford probably wouldn't be too judgmental about the auto industry's recent struggles because he was no stranger to debt himself. In 1899 the young mechanic and engineer started the Detroit Automobile Company with the backing of three prominent politicians, but hadn't quite mastered the innovation and production techniques that would eventually make him rich.

Ford quickly proved to be too much of a perfectionist, and his plant only produced 20 cars in two years as he painstakingly tinkered with designs. The enterprise went bankrupt in 1901 and reorganized into the Henry Ford Company later that year. Ford eventually left that group and finally got things right in 1903, when he founded the Ford Motor Company. Things didn't go so badly for the Henry Ford Company after he left, either; it changed its name into one you might find a bit more recognizable: the Cadillac Automobile Company.

Ford wasn't the only auto magnate who knew how bankruptcy felt, though. General Motors founder, William Crapo Durant, took a massive hit during the Great Depression that saw his fortune fall from $120 million to bankruptcy. He spent his last few years running a bowling alley in Flint, Mich.

3 Walt Disney

His name may be a stalwart brand today, but early in his career, Disney was just a struggling filmmaker with too many bills. In 1922, he started his first film company with a partner in Kansas City. The two men bought a used camera and made short advertising films and cartoons under the studio name Laugh-O-Gram. Disney even signed a deal with a New York company to distribute the films he was producing.

That arrangement didn't work out so well, though, as the distributor cheated Disney's studio. Without the distributor's cash, Disney couldn't cover his overhead, and his studio went bankrupt in 1923. He then left Kansas City for Hollywood, and after a series of increasingly successful creations, Disney debuted a new character named Mickey Mouse in 1928.

A shampoo containing real beer was marketed in the 1970s under the brand name Body On Tap.

Milton Hershey

Milton Hershey always knew he could make candy, but running a successful business seemed just out of his reach. Although he never had a formal education, Hershey spent four years apprenticing in a candy shop before striking out on his own in Philadelphia in 1876. Six years later, his shop went under, as did a subsequent attempt to peddle sweets in New York City.

Hershey then returned home to Lancaster, Penn., where he pioneered the use of fresh milk in caramel productions and founded the successful Lancaster Caramel Company. In 1900, he sold the caramel company for $1 million so he could focus on perfecting a milk chocolate formula. Once he finally nailed the recipe down, he was too rich (and too flush with delicious chocolate) for anyone to remember the flops of his early candy ventures.

5 Burt Reynolds

Burt Reynolds was one of Hollywood's biggest stars of the 1970s. Unfortunately, though, he spent money like his career would never hit a downswing. He owned mansions on both coasts, a helicopter, and a lavish Florida ranch. His financial situation became increasingly grim as he made boneheaded career choices and weathered a pricey divorce from Loni Anderson. By 1996, the Bandit owed $10 million to his creditors, and the royalties from *Cop and a Half* just weren't flowing in quickly enough. Reynolds filed Chapter 11 bankruptcy, from which he emerged in 1998.

Not only did he avoid having to sell his trademark mustache at auction to pay his bills, Reynolds even got to keep his Florida estate, Valhalla thanks to a homestead exemption that lets bankrupt people keep their homes. This exemption raised the ire of some observers who didn't think hanging on to a $2.5 million mansion while writing off $8 million in debt was quite in the spirit of bankruptcy laws. In fact, when the Senate passed measures tightening these loopholes in 2001, Reynolds' keeping his ranch was one of the examples they used to decry bankruptcy proceedings for being too easy on the wealthy. "There is no greater bankruptcy abuse than this," said Wisconsin senator Herb Kohl.

The long-running State Farm jingle ("Like a good neighbor, State Farm is there!") was written by Barry Manilow.

WHAT'S A HULU?

THE ORIGINS OF SIX HI-TECH NAMES

1 TiVo

Can you imagine if, instead of "TiVo-ing" the latest *Seinfeld* rerun, you were "Bongo-ing" it? "Bongo" and "Lasso" are just two of the 800 possible names marketing folks kicked around before settling on TiVo. The final name was cobbled together from "TV" and the engineering acronym "I/O," which stands for "input/output." Little did they know their noun would become a verb and their oddly named invention would forever change the way people watch television.

2 Bluetooth

Despite the lack of dignity displayed by people who shout into their Bluetooth headsets, the name of the device actually has a rather regal origin. In the 10th century, Danish King Harald Blatand was able to unite warring factions in Norway, Sweden, and Denmark under one banner. Similarly, the developers of the Bluetooth signal wanted to unite many different forms of technology—cars, computers, and mobile phones—under one communications network. Naturally, when they needed a name they went with the English translation of the Danish king's last name, "Bluetooth."

3 Hulu

Hulu means many things to many people. To some, it's a great online resource for watching their favorite TV shows and movies. To a native Hawaiian, however, it means "feather." To someone who speaks Swahili, it means "cease."

While these translations are accurate, the folks who named hulu.com took their inspiration from a couple of Mandarin Chinese definitions instead—"interactive recording" and "a hollowed-out gourd used to hold precious things." Despite this often misunderstood word, the website is rapidly becoming one of the biggest names in streaming video. Well, except in Indonesia . . .

When it was patented in 1970, the computer mouse was known by the catchy name "X-Y Position Indicator for a Display System."

4 BlackBerry

Would President Obama have fought so hard to keep his "LeapFrog" phone? LeapFrog was one of the names considered for the BlackBerry because the phone was leaps and bounds over everything else on the market. Another possibility was "Strawberry," because the tiny keys resembled seeds. When someone felt the word "straw" sounded too slow, another berry was suggested as a placeholder until something more official was developed. Over time, the temporary berry became so synonymous with the new device that it became the official name by default.

5 Nintendo Wii

Although the bathroom jokes almost write themselves, Nintendo had other ideas when they named their video game system. The word is pronounced "we," which emphasizes the social concept that Nintendo envisioned for the console. The name is also universal, without any direct translation into any particular language, reinforcing that all-inclusive idea and avoiding any Hulu-like confusion. Nintendo especially liked the double-i spelling because it looks like two people standing side-by-side. The name was not popular at first, but the concept obviously caught on, because Americans have purchased over 35 million Wiis since its debut in 2006.

6 Wikipedia

While the origin of the second half of the name might seem rather obvious, the first half is still a mystery to many. "Wiki" is used to describe any website content that is specifically designed to be edited by its users. Ward Cunningham first coined the term in 1994 to describe software that was meant to speed up the communications process between computer programmers. He borrowed the word from the Hawaiian language, where it means "fast," after hearing it in the Honolulu airport when an employee told him to take the "Wiki Wiki Shuttle" between terminals.

Many people mistakenly believe Wiki is an acronym for "What I Know Is." However, that definition was actually applied to the word after the fact, making it a backronym instead.

Bill Clinton sent only two emails as president. One was a test message to see if he was doing it right. The other was to John Glenn.

COMMERCIAL FAILURES

SIX AD CAMPAIGNS GONE TERRIBLY WRONG

Fowl Play: Perdue Farms in Mexico

Chicken magnate Frank Perdue built his family-owned farm into the fifth largest corporation in the industry through innovative marketing and darn good chicken. Things took an unpleasant turn and he ran for the border and took his product to Mexico, though.

The Perdue Farms' slogan was, "It takes a tough man to make a tender chicken," which, along with a photo of Perdue next to one of his birds, appeared on billboards throughout Mexico. Unfortunately, the catchphrase got a bit jumbled in the Spanish translation and became something more like, "It takes a sexually excited man to make a chicken affectionate," or "It takes a hard man to make a chicken aroused." Either way, Perdue ended up with egg on his face.

Speaking of poultry mishaps, when KFC tried out its slogan in Chinese, the Kentucky Fried tagline "finger-lickin' good" came out as "eat your fingers off."

2 You Say Potato: The Unsuspecting Guy in Miami

For every large corporation that's made a translation blunder, there are undoubtedly scores of smaller businesses that similarly messed up international opportunities. When Pope John Paul II visited Miami in September 1987, a rather enterprising gentleman thought he'd cash in by offering T-shirts with the phrase "I saw the Pope" in Spanish. Unfortunately, instead of using "el Papa" ("the Pope"), he mistakenly substituted "la Papa" ("the Potato"). And while spuds everywhere rejoiced at their newly found fame, all eyes were on the businessman, who found himself the subject of everlasting public ridicule.

3 How Ya Gonna Clean Up This Mess?: Electrolux in America

American companies aren't the only ones that have fumbled their ad campaigns on foreign soil. Sometimes, the embarrassment is imported to our very own shores. Case in point: Electrolux, the Scandinavian electronics company. Electrolux can make one heck of a refrigerator (Frigidaire), and if you need a vacuum cleaner that'll suck the chrome off a trailer hitch, they're your guys.

The phrase "always a bridesmaid, never a bride" was popularized in ads for Listerine mouthwash in the 1920s.

But the company ran into a little trouble trying to lure American consumers in the early 1970s. When the company took its catchy rhyming phrase "nothing sucks like an Electrolux" and brought it to America from English-speaking markets overseas, they failed to take into consideration the fact that "sucks" had become a derogatory word in the States. The serious (s)language barrier convinced the company to turn to a U.S.-based P.R. firm for future ad campaigns.

4 Water Isn't the Only Beverage to Avoid: Coors in Mexico

In the 1990s, Coors' famous beer tagline, "Turn It Loose!" targeted feisty, rad, outdoorsy types all over the world—except in Mexico. Despite hiring some pros to help hawk Coors Light to Hispanic customers, there was a really nasty mix-up. Details are sketchy, but it seems that the company chose the phrase *suéltate con Coors*, meaning "set yourself free with Coors." Somewhere along the line, though, advertisements appeared with *suéltalo con Coors*, which was translated to either "let it go loose with Coors" or "suffer from diarrhea with Coors." Either way, you're not gonna want to party with that action, dude.

5 Don't Judge the Jar by Its Label: Gerber in Africa

When Gerber first sold baby formula in Africa in the 1960s, it kept the same packaging it had created for the American market, which pictured a gorgeous, smiling, Caucasian baby beaming on the label. While this may seem the norm for baby food in the good ol' US of A, it wasn't the case in Africa. There, as Gerber execs soon discovered, companies routinely put pictures on the label of the actual contents that are inside—a result of the unfortunate reality that most consumers there are unable to read. Forget "Baby on Board"; we're talkin' Baby in Bottle (blended baby, no less). And no matter what language you speak, that's a recipe for disaster.

6 A Pun in the Oven: Parker Pens in Latin America

In 1935, the Parker Pen Co. invented and marketed a truly innovative product: a reliable fountain pen. Most businessmen of the day carried their pens in sparkling white shirt pockets, and the Parker model offered them the relief to be able to holster those puppies without worrying they would leak or stain. The pen was a wonder, and the ad slogan, "Avoid embarrassment, use Parker Pens," was a huge success. The next step? Go global, of course.

When they first expanded their market to Latin America, what the folks at Parker wanted to say was, "It won't leak in your pocket and embarrass you." Problem was that the Spanish word *embarazar* has a double-meaning; it means "to embarrass," but it also means "to impregnate." So, to some unsuspecting souls, the ad read: "It won't leak in your pocket and make you pregnant."

When Hostess released Hostess Turtle Pies for the 1991 release of *Teenage Mutant Ninja Turtles II*, the ad slogan read, "Fresh from the sewers to you!"

THE TAXMAN TAKETH

10 ARGUMENTS AGAINST PAYING YOUR TAXES (THAT WON'T WORK)

Don't want to pay your taxes? You'll have to get even more creative than these excuses, none of which held up in court.

1 "Taxes Are 'Voluntary.'"

This argument comes from a misunderstanding of the word "voluntary," which appears in a few tax-related sources, including the instructions that come with your 1040 tax form. Unfortunately, the legal definition of the word "voluntary" in this case refers to the process by which taxpayers report and pay taxes on voluntarily reported income, as opposed to a system where the government just tells you what to pay and you fork it over. Don't think that you can get tricky and say that while filing a tax return might be mandatory, actually paying the taxes is voluntary, either. The IRS already thought of that one, too.

Thomas Edison was an avid spokesman for concrete bedroom furniture. He held more than 40 patents directly related to concrete.

2 "Compensation Is Not Income."

Here's the argument: If you work for compensation, then you're not actually profiting. You're just bartering your time for money, which is a zero-sum transaction. Consequently, you haven't made a gain or profit that can be legally taxed. This transaction can be misconstrued as an "exchange," rather than income. The IRS dismissed this ploy as clever, but not convincing.

3 "Taxes in America Aren't for Americans."

Apparently there's a sentence or two in the tax code (which is more than 50,000 pages, by the way) that discriminates between U.S. and non-U.S. sources of income. It's just a small point explained so that folks don't pay double taxes if they happen to have income from multiple countries. A few individuals have plucked this one little idea and claimed that no taxes are due on income earned in America by Americans. By this logic, only aliens have to pay. The IRS rebuttal: "Read the other 49,999 pages and get back to us."

4 "Money Isn't Legal Tender."

Some folks are still a little peeved that they can't take a couple of Benjamins into their local banks and exchange them for equal amounts of silver or gold. They therefore claim that any income paid in such "worthless" tender cannot be taxed, as it inherently has no value. Truth is, they've got nothing to be peeved about. Article I, Section 10 of the Constitution says that the states cannot declare anything as legal tender other than gold and silver, but imposes no such limits on the Congress. So if you're paid in "worthless" Federal Reserve notes, you're welcome to donate them to the mental_floss Christmas party fund, but you still have to pay your taxes on them.

5 "I Am Not a Citizen."

Some creative former accountants and militia members got together and figured out that if they rejected their U.S. citizenship in favor of their state citizenship, they'd be outside of the tax-levying powers of the IRS. Or, put more succinctly, "I am a free-born citizen of (insert Mountain West state here), and you have no right to my money, Mr. Tax Man." The IRS rebuttal: Creative? Probably. Creepy and unconvincing? Definitely.

6 " 'The U.S.' Includes Only Federal Land."

Another state's rights argument claims that states are sovereign and only federal lands such as the District of Columbia, Guam, Puerto Rico, and federal enclaves like reservations and military bases are subject to federal taxation. The IRS rebuttal: Seriously? We've got a baby shower to go to in the third floor break room, and you're taking our time with this?

Each year British cities spend $222 million cleaning up discarded gum. Today all gum in the U.K. is taxed to defray the cost.

"Individuals Aren't People."

We're just going to quote the IRS on this one since it's pretty priceless: "Some maintain that they are not a 'person' as defined by the Internal Revenue Code, and thus not subject to the federal income tax laws. This argument is based on a tortured misreading of the Code."

What misreading you might ask? Well, the code defines a person as "an individual, trust, estate, partnership, or corporation." We can state without much reservation that we personally are not trusts, or even partnerships, but we'd have a hard time arguing that we are not individuals. This argument might work with your philosophy major pals, but it won't fly with the IRS.

8 "My Religion Doesn't Believe in Taxes."

Whether your religion doesn't like taxes or doesn't like the programs those taxes fund, the courts have held that "necessities of revenue collection through a sound tax system raise governmental interests sufficiently compelling to outweigh the free exercise rights of those who find the tax objectionable on bona fide religious grounds." Nice try, though.

9 "I Plead the Fifth."

This is a beautiful legal argument. "If I have income from illegal sources, then the reporting of such income forces me to incriminate myself in direct opposition to the rights granted me by the Fifth Amendment." However, the Supreme Court has established "that the self-incrimination privilege can be employed to protect the taxpayer from revealing the information as to an illegal source of income, but does not protect him from disclosing the amount of his income." Basically, you don't have to tell the IRS your income came from illegal iguana smuggling, but you still have to report the income.

10 "Taxes Are Slavery."

This argument asserts that the compelled compliance with federal tax laws is a form of servitude in violation of the Thirteenth Amendment. The short rebuttal: It isn't. The long rebuttal: It isn't, and that's insulting to millions of people descended from the people the Thirteenth Amendment was meant to protect.

The official currency of Ecuador is the United States dollar.

BIG DISCOUNTS, BIG PROBLEMS

SEVEN THINGS WALMART HAS BANNED

Think Walmart stocks just about everything? Not quite. Here are seven things Walmart has banned from its stores.

1 Barbie's Pregnant Pal

In 2002, Walmart cleared its shelves of Barbie's pregnant friend, Midge. The doll, which featured a removable stomach complete with deliverable baby, was part of Mattel's "Happy Family" set that also included her husband and son. However, customers complained about seeing pregnancy enter Barbie's universe, and Walmart pulled all of the Happy Family sets from its stores.

2 Suggestive Girls' Underwear

Someone at Walmart thought this would be a good idea: panties that say, "Who needs credit cards..." on the front and "When you have Santa" on the rear. The undergarments started showing up in Walmart's juniors departments in December 2007 and quickly started an Internet firestorm over the perceived message of using Kris Kringle as a sugar daddy. While the same joke would be fairly harmless on, say, a T-shirt, many women felt that its placement on underwear added a sinister sexual undertone aimed at adolescent girls. In response to the public outcry, Walmart pulled the offending underthings from its shelves.

TV's first commercial ran on July 1, 1941. The Bulova Watch Company paid $9 to air an ad before a Dodgers-Phillies game.

Workplace Romance

In November 2005, German courts ruled that Walmart could not ban all workplace romance at its German stores. The retailer had unsuccessfully tried to force all employees to sign off on a 28-page code of ethics that included prohibitions on "lustful glances and ambiguous jokes" and "sexually meaningful communication of any type."

Superbad DVDs

When the comedy *Superbad* hit store shelves in 2007, it came with a little extra: a replica of the fake Hawaii driver's license used by the self-dubbed "McLovin'." Most movie fans would simply see this freebie as a little reminder of one of the movie's funniest scenes, but Hawaiian authorities simply felt it was a fake ID. Honolulu mayor Mufi Hannemann requested that Walmart pull the DVD from store shelves across the state, and the retailer quickly complied.

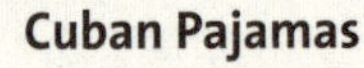

Cuban Pajamas

Walmart's Canadian stores found themselves in a pickle in 1997. The Canadian subsidiary had begun selling Cuban-made pajamas at eight bucks a pop, which enraged both the company's home office and the U.S. Treasury Department.

The stores quickly pulled the Cuban PJ's, which led to a second problem: This action may have violated a Canadian law that forbids abiding by the American embargo of Cuba. After the Ottawa government pointed out that Walmart could face a million-dollar fine for pulling the sleepwear from its shelves, the Canadian Walmarts reversed the ban after one week.

Music

Walmart has long declined to stock any music bearing a parental advisory warning for explicit lyrical content, but the company's sonic pickiness doesn't stop there. When the store carried Nirvana's album *In Utero*, it changed the song title "Rape Me" to the less offensive (and less coherent) "Waif Me." Similarly, the store declined to carry Prince's 1988 album *Lovesexy* because of a fairly tame cover that featured a nude photo of the artist.

A Shirt That Read "Someday a Woman Will Be President"

In 1995, a Miami-area Walmart pulled the progressive garment from its racks after consumer complaints. The shirt, which featured the character Margaret from *Dennis the Menace*, ran afoul of "the company's family values," so it went back to the stockrooms. Eventually more reasonable, non-Stone-Age heads prevailed, and the shirt made it back onto the shelves after three months in limbo.

The original Miss Chiquita Banana was drawn by Dik Browne, who created the comic strip "Hägar the Horrible."

HOLIDAY INS

FIVE BELOVED TRADITIONS INVENTED TO MAKE YOU BUY STUFF

Think you're impervious to advertisers' manipulations? Think again. Some of your favorite traditions might just be wildly successful viral marketing campaigns.

1 Green-Bean Casserole

America's favorite casserole dates back to 1955, when a chef named Dorcas Reilly created it for a cookbook designed to promote Campbell's products. By 2003, more than 20 million families (about one in four households) reportedly served the dish at Thanksgiving.

2 Rudolph the Red-Nosed Reindeer

The origin of Rudolph has nothing to do with Jesus or Santa. Instead, he sprang from the mind of Robert May, a copywriter for Chicago's Montgomery Ward department store. May wrote and illustrated the poem (that later became the song) for the store's holiday coloring

Rudolph was almost named either Rollo or Reginald. In the end May decided that Rollo sounded too happy and that Reginald sounded too British.

book in 1939. But Rudolph's fate was threatened when store execs realized that the animal's big, glowing honker might put off consumers because red noses were often associated with alcoholics. Luckily for May, shoppers embraced the story wholeheartedly. A whopping 2.4 million copies of *Rudolph the Red-Nosed Reindeer* were given out at the store that Christmas.

Diamond Engagement Rings

Prior to the 20th century, engagement rings were strictly luxury items, and they rarely contained diamonds. But in 1939, the De Beers diamond company changed all that when it hired ad agency N.W. Ayer & Son. The industry had taken a nosedive in the 1870s, after massive diamond deposits were discovered in South Africa. But the ad agency came to the rescue by introducing the diamond engagement ring and quietly spreading the trend through fashion magazines.

The rings didn't become de rigueur for marriage proposals until 1948, when the company launched the crafty "A Diamond is Forever" campaign. By sentimentalizing the gems, De Beers ensured that people wouldn't resell them, allowing the company to retain control of the market. In 1999, De Beers chairman Nicky Oppenheimer confessed, "Diamonds are intrinsically worthless, except for the deep psychological need they fill."

In addition to diamond engagement rings, De Beers also promoted surprise proposals. The company learned that when women were involved in the selection process, they picked cheaper rings. By encouraging surprise proposals, De Beers shifted the purchasing power to men, the less-cautious spenders.

4 Valentine's Day Candy

Greeting-card companies didn't invent valentines, but candy suppliers were very much behind the idea of giving out Valentine's Day candy. In fact, the tradition almost seems born out of jealousy. In 1892, *Confectioners' Journal* advocated persuading customers that candy was better than "cheap, grotesque" valentines. The floodgates flew open, and by 2004, consumers were buying more than 35 million heart-shaped boxes of candy each year.

5 Wedding Registries

In the 1900s, it was customary for only close family members to give wedding presents. But gradually, newlyweds came to expect gifts from friends, as well. Detecting a trend, department stores started to direct engaged customers to their home furnishings and kitchenware departments, encouraging them to think of their weddings as a time to acquire the tools for domestic life. In 1924, the Marshall Field & Company department store in Chicago created the first wedding registry, and the "tradition" took off. Today, up to 96 percent of American couples register their weddings.

Wilson Sporting Goods began as a way to use the slaughterhouse by-products of a meatpacking company.

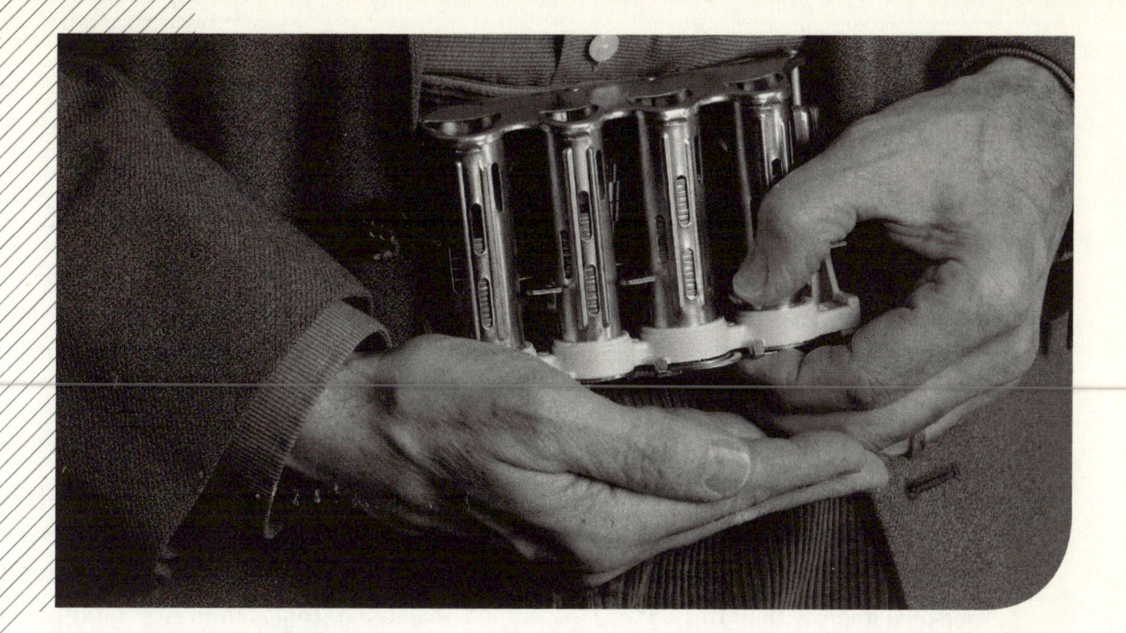

SERIOUS CHANGE

FIVE COINS THAT AREN'T BORING

The Stupidest Coin the Government Ever Made: The Racketeer Nickel

In 1883, the United States issued a newly designed five-cent piece called the "V nickel." The coin got its name because the value was indicated on the back simply with the Roman numeral "V," sans the word "cents." After all, it was obvious it was a nickel, right? Apparently not. Turns out, the V nickel was the same size as a U.S. $5 gold piece, and both coins featured a bust of Lady Liberty on the front.

It wasn't long before lightbulbs started going off in the heads of con men all across America. Within weeks of the V's debut, crooks were gold-plating the nickels and palming them off as $5 gold pieces. Despite the gold-plated nickels not looking like $5 coins and not being nearly as heavy, most people didn't notice, because the gold coins were rarely used in everyday purchases.

By April 1883, "gilded nickels" were both a national joke and a growing concern for commerce and law enforcement. Finally, embarrassed officials put an end to the scam by halting production of the nickels until new dies were prepared. This time, the redesigned backs read "V cents." Today, the V nickel remains a favorite among coin collectors.

If the average person received a penny for their thoughts, they'd rake in a cool $600 per day.

2 The Coin Your Mom Doesn't Want You to Pick Up: Leper Colony Coins

Leprosy, or Hansen's disease, was once among the most feared diseases in the world. Mistakenly believed to be highly contagious, it was a disfiguring and paralyzing condition that, until the 1900s, had no known cure. Sufferers were forced from their homes and exiled into colonies, where they wouldn't be able to spread the disease to the larger population.

Among attempts to quarantine lepers: Giving them their own currency. Many people feared leprosy could be transmitted by handling money, so special coins were minted (and, in some cases, paper bills printed) for leper colonies in areas including Venezuela, Brazil, Colombia, the U.S. Canal Zone, and the Philippines. Some city officials found another convenient use for leper money—paying inmates for their work and allowing them to buy personal items with it. This, so the logic went, prevented prisoners from ever being able to save up "real" money to aid in an escape.

3 The Coin You Can Never Take on an Airplane: Spanish Pieces of Eight

In the New World, colonists had to get creative when it came to currency. Because the British were too cheap to mint coins for their American settlements, colonists had to make do with barter, paper money, or whatever foreign coins they could scrape up through trade. Fortunately, Spain's New World colonies were rich in silver mines, and the Spanish had plenty of coins to toss around.

At the time, Spain minted coins about the same size as the Germanic silver thaler coins of Europe, and Americans took to calling them "Spanish dollars." But officially, Spanish dollars were valued at eight reals (*real* being Spanish for "royal"). So how do you make change for a Spanish dollar? For our colonial forefathers, it was easy. Knowing that silver is a fairly soft metal, they'd just take a mallet and a chisel, or even an ax, and slice up the coin like a pizza. The cut slices were called "bits," or pieces of eight. A 2-real piece was worth about 25 U.S. cents, which is why a quarter is sometimes referred to as "two bits." Another term for cut coin slices was "sharp silver," because the points were indeed sharp enough to cut cloth or even skin.

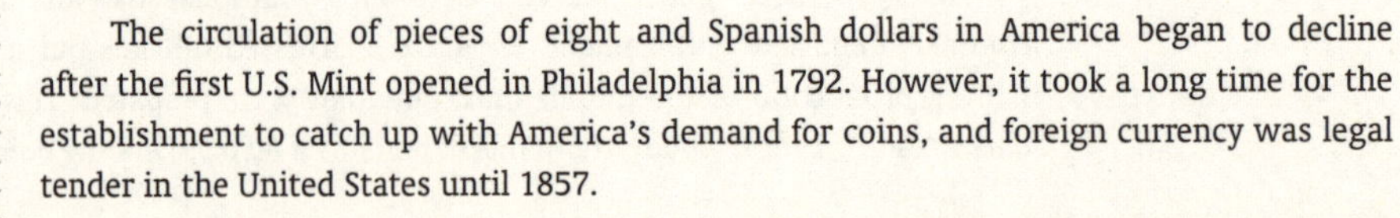

The circulation of pieces of eight and Spanish dollars in America began to decline after the first U.S. Mint opened in Philadelphia in 1792. However, it took a long time for the establishment to catch up with America's demand for coins, and foreign currency was legal tender in the United States until 1857.

Although pennies are worth less than their component metals, federal law prohibits melting them down.

4 The Not-Quite Counterfeit Coin: The 1804 Silver Dollar

America's most famous rare coin is the 1804 silver dollar. Why so special? Because it was actually made by mistake. Due to governmental budget constraints, the production of silver dollars was halted in the early 19th century. And while a few thousand $1 coins were minted in 1804, they were produced frugally, using the previous year's dies. Ironically, the first $1 coins dated 1804 weren't made until 1834, when the United States decided to present the King of Siam and the Sultan of Muscat with a diplomatic gift: complete sets of American coins. Records at the U.S. Mint correctly listed 1804 as the last year silver dollars were made, but didn't specify that the last ones were dated 1803. Consequently, American officials decided to strike a few new dollars with the date 1804, and ended up creating a coin that had never before existed.

Today, there are only 15 of these 1804 silver dollars left. Eight of them were from the batch minted as diplomatic gifts. The other seven were produced between 1858 and 1860, when an employee of the Philadelphia Mint decided to get rich quick on the coin collector's market. Using the mint's silver and equipment, he struck a number of new 1804 silver dollars to sell to collectors. The phony coins (although illegally produced, they're technically not counterfeits because they were made at a U.S. Mint) were eventually found and melted down—all but seven of them, that is. One of these restrikes was auctioned in 2003 for $1.21 million, but that's chump change compared to the $4.14 million paid for one of the original coins back in 1999.

5 The "Choose Your Own Coin" Coin: Blank Coins

The quality-control regulators at our mints do a great job of catching mistakes, but luckily for collectors, some botched coins do make their way into circulation. Among the more common errors are blank coins, such as a one-cent piece. Coins are made by pressing a die onto a planchet, or coin blank, that's been punched out of a piece of sheet metal. Sometimes, a planchet slips through the process without being struck, and a blank coin ends up in an otherwise ordinary roll of pennies. Other common errors include coins struck off-center, coins struck on the wrong planchet (i.e., the image of a quarter stamped onto a penny), and double-struck coins.

At one point, the U.S. Mint was spending $35,000 a year to store 122 million unneeded Susan B. Anthony dollar coins.

YOU'VE GOTTA START SOMEWHERE

STRANGE EARLY JOBS OF 13 FAMOUS PEOPLE

1. Need a rat catcher who can act? Call Warren Beatty. He caught rodents to pay the bills before hitting it big.

2. When actress Amy Adams needed cash to buy a car after high school, she spent two months working as a Hooters girl.

3. Rod Stewart had a number of jobs before his music career took off, but grave digger was undoubtedly the creepiest.

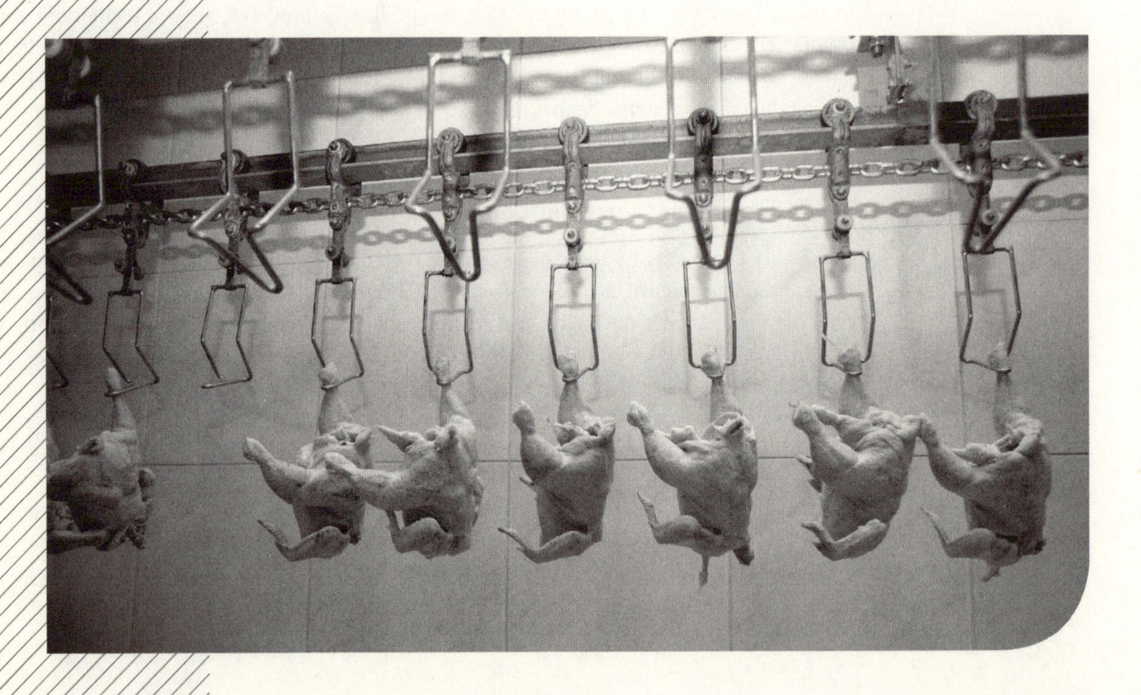

4. You never know how a weird job might pay off. Singer Chubby Checker had a job as a chicken plucker at Fresh Farm Poultry in Philly. His boss would let Checker sing to entertain the customers. When the boss realized just how talented his chicken plucker was, he arranged for Checker to have a recording session with Dick Clark.

Wendy's founder, Dave Thomas, dropped out of high school but returned to get his GED 45 years later. His GED class voted him Most Likely to Succeed.

Before Stephen King became the world's leading master of horror he worked in an industrial Laundromat.

Brad Pitt did all sorts of things to earn a buck while he tried to start his acting career, including dressing as a giant chicken to promote an el Pollo Loco restaurant.

Writer Jack London spent time working as a poaching oyster pirate. His sloop had an awesome name: *Razzle-Dazzle*.

8 When Christopher Walken was a kid, he joined the circus. He took an unpaid job with a small act and even did a little lion taming, although he later claimed the lion was very old and "really more like a dog."

9 Of course, some early jobs predict the future rather well. Before rising to prominence with Black Sabbath, Ozzy Osbourne worked in a slaughterhouse.

10 Author R. L. Stine of the *Goosebumps* horror series wrote some less chilling words for children, too—prior to his horror series, Stine was writing joke books for kids under the name Jovial Bob Stine. He was also the cocreator and head writer for the Nickelodeon show *Eureeka's Castle*.

11 Bram Stoker of *Dracula* fame started writing as a theater critic for the *Dublin Evening Mail* and also wrote a thrilling piece of nonfiction entitled *The Duties of Clerks of Petty Sessions in Ireland*.

12 *Slaughterhouse-Five* author Kurt Vonnegut was the manager of a Saab dealership in West Barnstable, Massachusetts—one of the first Saab dealerships in the United States. He also worked in public relations for General Electric, and was a volunteer firefighter for the Alplaus Volunteer Fire Department.

13 This one's our favorite: Novelist Harper Lee had worked as a reservation clerk at Eastern Airlines for eight years when she received a note that read, "You have one year off from your job to write whatever you please. Merry Christmas." By the next year, she'd penned *To Kill a Mockingbird*.

Dr. Ruth was trained as a sniper by the Israeli military.

10 POP CULTURE LISTS TO BREAK OUT ON THE RED CARPET

TITLE FIGHTS

WHAT SEVEN OF YOUR FAVORITE MOVIES WERE ALMOST CALLED

Back to the Future

During the filming of *Back to the Future*, Universal Studios honcho Sid Sheinberg fired off a memo to all involved in the production stating that no movie with the word "future" in the title had ever succeeded at the box office. He suggested that the name of this project be changed to *Spaceman from Pluto*. According to writer/producer Bob Gale, Steven Spielberg "earned his executive producer fee" by stepping in and sending back a note that thanked the studio head for his joke memo.

Sound designer Ben Burtt created R2D2's voice by whistling and making fart noises into an old tape recorder.

Pretty Woman

Pretty Woman was based on a script written by J. F. Lawton called *3000* (the amount of money paid for a week's worth of the hooker's "company"). In the original story, Julia Roberts' character was not only a prostitute, she was also a crack addict. When Disney bought the script, they lightened up the script and decided that *3000* didn't work. They picked a new title that luckily allowed the producers to use Roy Orbison's signature tune to accompany the obligatory Julia Roberts-trying-on-clothes montage.

3 *Tootsie*

Would I Lie to You? went through at least a half-dozen rewrites by as many writers before it finally hit the big screen in 1982 as *Tootsie*. Star Dustin Hoffman suggested the new title, which was also the name of his mother's dog.

Boys Don't Cry

Hilary Swank nearly won an Academy Award for *Take It Like a Man*. Eventually the intense flick took its title from a signature song of one of the darkest bands in rock, The Cure.

Help!

The Beatles called their second feature film *Help* during the first few weeks of production, but things got sticky when director Richard Lester learned that another flick had already registered that title. The Fab Four's movie then became known as *Eight Arms to Hold You*. No one cared for that name, but luckily Lester found out that by adding an exclamation point to the word "Help" he could skirt the copyright laws.

Annie Hall

Anhedonia is the scientific term for the inability to experience pleasure, and up until the opening credits were finally filmed, it was the name Woody Allen had in mind for his 1977 "serious comedy." United Artists insisted that such an unmarketable title would doom the film at the box office, and Allen relented by changing the title to *Annie Hall*.

Blazing Saddles

Script writer Andrew Bergman originally called his 1974 Western spoof (and the lead character) *Tex X*, as a sly nod to Muslim leader Malcolm X. Director Mel Brooks was never enamored with that title, though. One morning when Brooks was taking a shower, the words "blazing saddles" popped into his head. Considering the classic bean scene, the new title seemed apropos.

The first person to use the word "period" in a U.S. TV commercial was future *Friends* star Courteney Cox, in a 1985 Tampax ad.

MORE THAN A PRETTY FACE

FOUR ACTORS WHO CHANGED HISTORY

Lola Montez—The Actress Who Ended an Empire

Few 19th-century entertainers were as colorful as Irish actress and dancer Lola Montez (born Eliza Rosanna Gilbert), who appeared in Broadway shows and performed around Europe as a "Spanish" dancer. She was also banished from Warsaw for publicly criticizing the ruling despot, attacked a newspaper editor in Australia when he published a bad review of her show, gossiped with the Tsar, and was the lover of composer Franz Liszt.

In her most notorious episode, Montez became mistress of King Ludwig I of Bavaria in 1846, despite a 35-year age difference. Her fierce temper and arrogance made her very unpopular with his subjects, who were furious to see the influence she held over the famously amorous king, especially when he made her Countess of Landsfeld. Noble or not, her scandalous behavior contributed to the fall from grace of the popular king (who had ruled for 22 years), inspiring thoughts of revolution. Ludwig was forced to abdicate in 1848. Montez died in 1861 at age 39.

2 Florence Lawrence—Auto Pioneer

Canadian-born Florence Lawrence is mostly forgotten, but devoted film buffs remember her as the world's first movie star. Initially, like all movie actors in the early days, she was uncredited for her work, and known to her fans as "the Biograph girl" (named after the studio that made her films). Though her face became famous, her name was unknown until Hollywood mogul Carl Laemmle brought her to his new IMP Company in 1910, revealing her name in big newspaper ads. At her first public appearance, in St. Louis, the crowds she drew were bigger than those that had greeted President Taft the previous week.

In her spare time, however, Lawrence was a tinkerer. The daughter of inventors, she was one of the first car owners and a true automobile geek, inventing such accessories as the "auto signaling arm," a forerunner of the turn signal (she placed an arm on the back of the fender that could be activated with the push of a button) and the automatic "full stop" sign, an early version of brake lights.

Sadly, her genius didn't extend to filling out patent forms, so others became rich by enhancing her inventions. Nor did she realize that popular film stars could demand ridiculously high salaries. She eventually died in poverty in 1938.

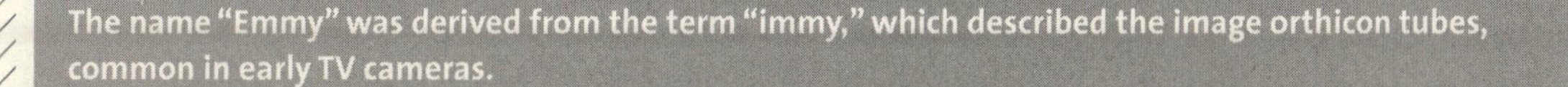
The name "Emmy" was derived from the term "immy," which described the image orthicon tubes, common in early TV cameras.

Hedy Lamarr—Mastermind of the Wireless

There must be something about movie stars and inventiveness. Hedy Keisler made history with her teenage nudity in the daring Czechoslovakian film *Ecstasy* (1932), and won even greater fame when she moved to Hollywood and took the name Hedy Lamarr. In films like *Algiers* (1938) and *Samson and Delilah* (1949), she was known as one of the most sultry and beautiful women in the movies.

But Lamarr had brains as well as beauty. During World War II, she invented a radio guidance system for torpedoes, which she developed with the help of another clever Hollywood friend, composer George Antheil. Known as "frequency hopping," it consisted of two synchronized pianola rolls, allowing technicians to switch control frequencies so the torpedo could escape enemy tracking. Though they received a patent in 1942, the War Department declined to use it. It was later adapted for satellite communications, and is now widely used in cellular phones and other modern technology. As the patent had expired before most of this, neither Lamarr nor Antheil profited from their cleverness, but Lamarr was given due recognition before her death in 2000.

4 Kevin Bacon and His (Social) Network Solutions

Well, sort of. Remember the 1990s party game Six Degrees of Kevin Bacon, in which players try to link any given movie actor, through a short chain of films, to Kevin Bacon, the star of *Footloose*? Bacon once commented in an interview that he'd worked with everyone in Hollywood, and three Albright College students created and popularized the game.

Of course, the idea of networking (getting a job from the friend of a friend) was nothing new, but around the time Six Degrees of Kevin Bacon originated, scientists Duncan Watts and Steve Strogatz were exploring the world of network theory, a new field of scientific research suggesting that large groups (be they viruses or football crowds) don't connect randomly, but are structured around nodes. The theory was all well and good, but they needed to prove it by studying some real networks. Strangely, few networks had been mapped—but then they discovered The Oracle of Bacon, a cheat-sheet website for the Kevin Bacon game, created by student Brett Tjaden and linked to the Internet Movie Database.

Tjaden's program, in the hands of the scientists, advanced the research considerably. Network science principles have already led directly to the capture of Saddam Hussein, as the military moved through the dictator's social networks. But there are hub networks everywhere from computer chips to human cells. In the future, it is hoped, scientists will map out networks to combat terrorism, predict pandemics, even cure cancer. When this happens, remember to thank a certain Hollywood star for an offhand comment he once made in an interview.

In the 1930s, Lloyd's insured Shirley Temple for $25K with two conditions: She couldn't get hurt while drunk or take up arms during war.

CELEBRITY NAME-CALLING

THE ORIGINS OF 10 CELEBRITY STAGE NAMES

1 Whoopi Goldberg

Whoopi Goldberg took her stage name from the whoopee cushion. The actress, who was born Caryn Johnson, said that she was working in a theater in San Diego with small dressing rooms when she had a bit of a problem with gas. Goldberg would occasionally break wind during costume changes in the cramped space, and castmates would accuse her of being "like a whoopee cushion."

According to Goldberg, she considered going by the name "Whoopi Cushion" when her comedy career took off, but her mother argued that nobody would take her seriously with such a silly name. Her mom thought it would be smarter to pair "Whoopi" with a more serious name and proposed that her daughter use "Goldberg."

2 Albert Brooks

Albert Brooks is a funny man, but he probably wouldn't have made it too far in show business with his birth name: Albert Einstein. Brooks originally tried to go by his first and middle names, Albert Lawrence, but decided that combo "sounded like a Vegas singer." The name Brooks was already in his family, so he ran with that. His brother, Bob Einstein, actually kept the family surname when he entered showbiz, but even he's better known by an alias: Super Dave Osborne.

3 M. C. Hammer

M. C. Hammer got his nickname from his childhood job with the Oakland Athletics. Eccentric A's owner Charlie O. Finley loved Stanley Kirk Burrell, the talented kid who danced in the team's parking lot and eventually became a batboy and an errand boy for the club, and the benevolent owner called him "Little Hammer" because he thought Burrell looked like "Hammerin'" Hank Aaron. When the Little Hammer picked up the mic, he became M. C. Hammer.

4 Harry Houdini

Harry Houdini was born Ehrich Weiss, but he took on the stage name Harry Houdini as a tribute to famed French magician Jean Eugene Robert-Houdin. The "Harry" name was simply an American version of his childhood nickname "Ehrie."

John Matuszak, who played Sloth in *The Goonies,* was the first overall pick of the 1973 NFL draft as a defensive end.

Iron Eyes Cody

Iron Eyes Cody was one of Hollywood's most beloved Native American actors throughout the 20th century; you might remember him as the "Crying Indian" in the famous "Keep America Beautiful" ads. One thing most audiences didn't know, though, was that Cody wasn't actually Native American. For most of his life, though, he hid his Sicilian heritage and maintained that he was actually part Cree and Cherokee. He even married a Native American woman. This arrangement surely made it easier to land Native American roles than his real name, Espera "Oscar" de Corti, would have.

Snoop Dogg

Snoop Dogg was born Calvin Broadus, but his parents nicknamed him "Snoopy" because he looked like the famous cartoon beagle. Nothing is quite as gangsta as hanging out with Woodstock and Charlie Brown.

Jackie Chan

The fearless martial artist was working at a construction site in Australia when he got his famous nickname. One of his fellow workers couldn't pronounce Chan's first name, Kong-sang, so he referred to Chan as "little Jack." The name soon morphed into "Jackie," and eventually it stuck.

The title of Aerosmith's "Walk This Way" was taken from a line in Mel Brooks' comedy *Young Frankenstein*.

8 Sid Vicious

Sid Vicious got his famous stage name from Sex Pistols frontman John "Johnny Rotten" Lydon's old pet hamster, Sid. The singer's father was playing with Lydon's hamster one day when the rodent bit him and forced him to exclaim, "Your Sid is vicious!" Lydon thought the remark was so amusing that he started calling his friend "Sid Vicious."

9 LL Cool J

LL Cool J stands for "Ladies Love Cool James," as you may already know. What you might not know, though, is that the name wasn't necessarily true when it made its debut. When James Todd Smith and a buddy were 16, they began calling themselves LL Cool J and Playboy Mikey D in the hopes that it would help their stock with the girls they tried to woo. In a 2008 interview with CBS' *Early Show*, LL Cool J admitted that the ladies didn't actually love Cool James quite yet, saying, "It was just wishful thinking, just hoping for the best."

Alice Cooper

Alice Cooper got his name from a Ouija board. The shock rocker, who was born Vincent Furnier, was supposedly playing with a Ouija board in the late 1960s when a 16th-century witch doctor named Alice Cooper contacted him. Furnier and his buddies then started a band called Alice Cooper with the magnetic Furnier in the lead role of "Alice." Since the name originally referred to the whole band and not just Furnier, he continues to pay an annual royalty to his old bandmates for the commercial use of the Alice Cooper name.

Fear Factors: Eight Celebs and Their Incredibly Odd Phobias

1. **Billy Bob Thornton**—Antiques
2. **Alfred Hitchcock**—Eggs, specifically runny ones
3. **Christina Ricci**—Houseplants
4. **Tyra Banks**—Dolphins
5. **Matthew McConaughey**—Revolving doors
6. **Nicole Kidman**—Butterflies
7. **Oprah**—Chewing gum
8. **Madonna**—Thunder

Technicians created the sound effects for the famous shower scene in Hitchcock's *Psycho* by repeatedly stabbing a casaba melon.

ROMANIA SHOOTS J.R.

FOUR TV SHOWS THAT CHANGED HISTORY

Dallas—The Show That Overthrew a Dictator

Dallas was one of the most popular TV shows in history—and nowhere was it more talked about than in Nicolae Ceausescu's communist Romania. How did the soap opera get past Romanian censors? With help from Dallas leading man, J.R. Ewing, of course. Because J.R. was portrayed as a despicable oil baron, Ceausescu's government presumably decided the show must be anticapitalist.

Whatever the reasoning, *Dallas* became a runaway hit when it arrived in Romania in 1979. A series about wealthy, beautiful people (evil or not) was an inspiration to Romania's poor and dejected masses. Eventually, the government decided such Western television was a bad influence, and Dallas was taken off the air in 1981.

But by then, it was too late. The fantasies of Western life lived on in the imaginations of Romanians, and in 1989, Ceausescu was overthrown during a public uprising. Not incidentally, the actor who played J.R., Larry Hagman, visited Romania some years later and was treated as a hero. In an interview following the experience, Hagman said, "People from Bucharest came up to me in the street with tears in their eyes saying, 'J.R. saved our country.' "

General Electric Theater—The Show That Turned Ronald Reagan into a Republican

In the early 1950s, film actor Ronald Reagan was at a low point in his career. So when Taft Schreiber, of the Music Corporation of America, got him a gig as the host of the anthology series *General Electric Theater*, Reagan jumped at the opportunity. Reagan not only hosted the show, but also toured America as a "goodwill ambassador" for the electricity giant, giving speeches to plant employees and acting as its public spokesperson.

By the time *General Electric Theater* was canceled in 1962, Reagan was a new man. Turns out, all those years defending free enterprise for one of the nation's biggest multinational companies had transformed Reagan into one of America's leading conservative speakers. Although the actor had long been a Democrat, the Republican Schreiber convinced Reagan to change political parties. Four years later, the newly Republican Reagan was elected governor of California, and the rest is presidential history.

Dr. Sam Sheppard, the inspiration for *The Fugitive*, became a pro wrestler after he was cleared of his wife's murder.

3 *Star Trek*—The Show That Designed the Future (of Society)

Defying all stereotypes, the heroic crew of the U.S.S. *Enterprise* was comprised of a mix of races—and among them were some high-ranking women. Here again, *Star Trek* became an inspiration—only this time, to minorities and women, rather than tech junkies. Lieutenant Uhura, played by African-American jazz singer Nichelle Nichols, showed audiences that black women could be senior officers and hold positions of power. In fact, when Nichols contemplated quitting the series during its first year, she was persuaded to keep the role by none other than Dr. Martin Luther King, Jr., who said, "Don't you realize how important your character is?"

Years later, women ranging from Whoopi Goldberg to Dr. Mae Jemison, the first African-American female astronaut, cited Lieutenant Uhura as a major inspiration in their careers. Nichols even spent time working for NASA on an astronaut-recruitment program—an initiative that roped in such people as Sally Ride and Guy Bluford, the first American woman and African-American man in space, respectively.

4 *The Smothers Brothers Comedy Hour*—The Show That Swung an Election

The Smothers Brothers Comedy Hour was many things. It was the first network TV show to make fun of the Establishment, support America's counterculture, and have enough nerve to put blacklisted singers (such as Joan Baez and Pete Seeger) back on the air. Ironically, however, the show's major achievement might have been making Richard Nixon president.

As a gag, show star Pat Paulsen ran for office during the 1968 presidential election. "I'm consistently vague on the issues," announced Paulsen on national television, "and I'm continuing to make promises that I'll be unable to fulfill." Regardless of his humorous motives, Paulsen seemed to have a "Ralph Nader Effect," stealing 200,000 votes from the Democrats and helping to swing one of the closest elections in history. Nixon narrowly defeated Democratic candidate Hubert Humphrey. "Hubert Humphrey told me I cost him the election," recalled Paulsen, "and he wasn't smiling when he said it."

Shari Lewis, the puppeteer behind Lamb Chop, cowrote an episode of the original *Star Trek* TV series with her husband.

CANCELED!

10 TV SHOWS SCRAPPED AFTER JUST ONE EPISODE

1. ***Heil Honey I'm Home!***

   Can you imagine a worse premise for a sitcom than the escapades of Hitler and Eva Braun? No? Well, picture this: a Jewish couple moves in next door. Seriously.

2. ***The Will***

   Canceled in 2005 because apparently people didn't find family members and friends competing to be named the beneficiary of a loved one's will too palatable.

3. ***Who's Your Daddy?***

   This 2005 reality show asked an adopted woman to pick her biological father out of a group of phonies. Due to huge backlash, the show ended up being a "special" instead of a series premiere.

When *The Dukes of Hazzard* was in its heyday, the General Lee—the car—received over 30,000 pieces of fan mail each month.

❹ *Australia's Naughtiest Home Videos*

This one's notable because it was actually canceled while it was on the air. The owner of the network was so horrified that he called the station and ordered it pulled immediately. The network cut to a *Cheers* rerun.

❺ *Beware of Dog*

Capitalizing on the *Look Who's Talking* trend about 10 years too late, this 2002 show featured the inner thoughts of a dog adopted by some suburbanites.

❻ *Comedians Unleashed*

Another *Animal Planet* strikeout—a stand-up comedy show with animal-themed jokes. It was a dog.

❼ *Emily's Reasons Why Not*

Poor Heather Graham. ABC's programming chief canceled her starring vehicle the day after it aired because he decided that it wasn't going to get any better.

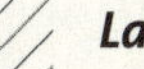

❽ *Lawless*

In 1997, retired Seattle Seahawks linebacker Brian Bosworth tried his hand at acting. His private investigator character couldn't solve the mystery of how to get to a second episode.

❾ *The Melting Pot*

"Mr. Van Gogh" is an illegal Pakistani immigrant in London. Hilarity ensues. Or not.

❿ *Secret Talents of the Stars*

Sadly, no one cared about Danny Bonaduce riding a unicycle, Marla Maples doing gymnastics, or Prince's former drummer Sheila E. juggling. So, the rest of the stars' secret talents remained hidden.

The Dukes of Hazzard used 229 different Dodge Chargers as the General Lee. Of those, 210 were destroyed during the course of the series.

GIG MISTAKES

FIVE FAMOUS ACTORS AND THE ROLES THEY TURNED DOWN

Sean Connery wasn't supposed to be James Bond. Keanu wasn't supposed to be "the One." So, who were the original choices? Here are five actors and the legendary roles they turned down.

1 THE ROLE: James Bond in *Dr. No*

WHO LET IT GET AWAY: Cary Grant. Despite being Bond producer Albert Broccoli's best man, Grant said, "I don't" to the offer, and Sean Connery got the role instead. Of course, many studio executives objected to the decision, and even Bond creator, Ian Fleming, said Connery "wasn't exactly what I had in mind."

REGRETTABILITY METER: Low. By the 1960s, Cary Grant already had a spectacular film career. If he'd accepted the role (as Broccoli later revealed), it would've been just a one-movie deal.

Gary Cooper turned down the role of Rhett Butler in *Gone with the Wind*, saying, "I'm just glad it will be Clark Gable falling on his face and not Gary Cooper."

❷ **THE ROLE: Neo in *The Matrix***

WHO LET IT GET AWAY: Will Smith turned it down to star in the forgettable action flick *Wild Wild West*, and the part went to Keanu Reeves.
REGRETTABILITY METER: Low. In an interview with *Wired*, Smith said, "I would have absolutely messed up *The Matrix*. At that point I wasn't smart enough as an actor to let the movie be—whereas Keanu was."

❸ **THE ROLE: Vincent Vega in *Pulp Fiction***

WHO LET IT GET AWAY: Michael Madsen, who was stuck in lengthy rehearsals for *Wyatt Earp*. John Travolta got the role instead and, almost overnight, transformed from a Hollywood has-been into one of the most bankable stars in the business.
REGRETTABILITY METER: High. Madsen called *Wyatt Earp* "a big waste of time."

❹ **THE ROLE: Gandalf in the *Lord of the Rings* Trilogy**

WHO LET IT GET AWAY: Sean Connery, who'd never read the J.R.R. Tolkien series and claimed he "didn't understand the script." (Can you say, "Karma"?)
REGRETTABILITY METER: High. In return for playing the role, New Line Cinema offered the Scottish actor up to 15 percent of worldwide box office receipts, which would have earned Connery more than any actor had ever been paid for a single role—as much as $400 million.

❺ **THE ROLES: Sundance in *Butch Cassidy and the Sundance Kid*, Jimmy "Popeye" Doyle in *The French Connection*, and Captain Benjamin Willard in *Apocalypse Now*.**

WHO LET THEM GET AWAY: Steve McQueen.
REGRETTABILITY METER: Tragically high. McQueen turned down the role of Sundance simply because costar Paul Newman refused to give him top billing. Later, McQueen declined the lead in *The French Connection* because he felt the part was too similar to the tough cop he'd played in 1968's *Bullitt*. Gene Hackman took the part and won an Oscar for it. And finally, in 1978, McQueen told *Apocalypse Now* director Francis Ford Coppola to shove off when he was offered the lead. McQueen's nonnegotiable asking price was $3 million; plus, he didn't feel like spending four months shooting in the Philippines jungle.

The Nigerian film industry, or "Nollywood," has replaced Hollywood as the world's second biggest producer of movies. (India's Bollywood makes the most.)

EIGHT CELEBRITY INVENTORS WHO HOLD PATENTS

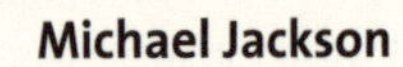

1 Michael Jackson

How did Michael Jackson seemingly lean in defiance of gravity in the video for "Smooth Criminal"? He wore a pair of specially designed shoes that could hitch into a device hidden beneath the stage. Jackson and two coinventors received a patent for the "method and means for creating anti-gravity illusion" in 1993.

2 Abraham Lincoln

Honest Abe held quite a few jobs before becoming a politician, and in one of these gigs he helped float a boatload of goods down the Mississippi River. At one point, the boat got stuck in a shallow spot, and it took quite a bit of effort to wrench it free. Lincoln thought that there must be a better way to keep ships off of shoals, so he invented a convoluted device that placed a set of bellows on the bottom of a boat. Lincoln's reasoning was that if the boat got in a sticky situation, sailors could fill the bellows with air to make the ship more buoyant.

Lincoln received Patent Number 6469 for this invention in 1849, but unfortunately, the creation never made it into stores. It turned out that all of the extra weight associated with adding the bellows device to a ship actually made it *more* likely that the boat would get stuck.

3 Eddie Van Halen

Part of guitar wizard Eddie Van Halen's signature sound is his two-handed tapping technique, but letting all 10 fingers fly while simultaneously holding up the guitar's neck could get a bit tricky. Van Halen came up with a novel way to get around this problem, though; he invented a support that could flip out of the back of his guitar's body to raise and stabilize the fretboard so he could tap out searing songs like "Eruption." While Van Halen was obviously interested in improving his guitar work, the patent application he filed in 1985 notes that the device would work with any stringed instrument. Want to tap out a scorching mandolin solo? Find someone selling Eddie's device.

The bottle that served as Jeannie's home in *I Dream of Jeannie* was actually a painted Jim Beam's Choice decanter.

Jamie Lee Curtis

In 1987, Curtis designed and patented a disposable diaper that included a waterproof pocket that held baby wipes. She hasn't profited from her idea yet, though, since she refuses to license the patent until diaper companies make biodegradable products.

Penn Jillette

In 1999, everyone's favorite wisecracking illusionist received a patent for a "hydro-therapeutic stimulator." What exactly does that mean? According to the application, it's "a spa of a type including a tub for holding water and a user, in particular, a female user." The spa's jets are strategically located to make the experience a bit more, ah, enjoyable for female bathers.

Marlon Brando

To say Brando got a bit eccentric in his golden years is something of an understatement, but the aging actor also started to get innovative. Brando's inventiveness focused on the drums, and in 2002 he received a patent for a "drumhead tensioning device and method," one of several patents he held for drum devices.

Lawrence Welk

Your grandma's favorite accordionist and bandleader was also an inventor. In 1953, Welk received a design patent for a new type of ashtray that looked like (what else?) an accordion. Not a huge breakthrough for humanity, but it went nicely with Welk's other patent; 10 years earlier he had received a design patent for a menu card that looked like a singing chicken.

Steve McQueen

McQueen's driving abilities extended far beyond his legendary chase scenes in *The Great Escape* and *Bullitt*. In fact, he was a pretty serious motorcycle and car racer who toyed with the idea of someday becoming a professional racer. He even competed in some big-name races, like the prestigious 12 Hours of Sebring. McQueen didn't just drive his cars, though; he also liked to tinker with them. In 1969, he filed a design patent for an improved bucket seat, and that's how he became the proud owner of patent number D219584.

Magician Teller of Penn & Teller fame holds one of the few one-name U.S. passports in existence.

POP TORTS

THREE ODDBALL LAWSUITS PLUCKED FROM POP CULTURE

❶ Those Freeloading Girl Scouts Stop Singing

When an artist holds the copyright to a work, they don't just have the exclusive right to record it. They also own the sole rights to broadcast or perform the song in public. Any band or group performing a cover version of a song owes royalties to the copyright holder.

This system is what got those adorable Girl Scouts in trouble. In 1995, ASCAP, one of the two major companies that collects royalties for copyright holders, decided that summer camps were getting away with publicly performing copyrighted campfire songs without paying any licensing royalties. ASCAP decided to derail the gravy train and ask these camps for the cash.

From a legal standpoint, ASCAP was within its rights. Requesting that nonprofit camp directors pay annual fees of as much as $1,400 or face six-figure fines or a year in prison isn't the world's greatest P.R. move. Girl Scout camps were hit particularly hard, and TV reports and a major story in *The Wall Street Journal* recounted tales of young lasses having to learn the Macarena in silence.

At one point the camps became so fearful of legal retribution that they stopped allowing the girls to sing "Happy Birthday" to one another lest a camp director be forced to serve hard time. As the P.R. nightmare for ASCAP worsened, the copyright holders finally relented. ASCAP now charges the Girl Scouts $1 a year to license its portfolio, a symbolic compromise that reasserts the group's ability to demand these kind of fees.

❷ Trivia Guru Sues for a Nontrivial Amount

When writer Fred L. Worth began playing Trivial Pursuit in the early 1980s, he noticed something odd. The questions felt very familiar. Worth became suspicious that some of the facts from his book *The Complete Unabridged Super Trivia Encyclopedia* had been used for the game's cards. A closer inspection revealed that even some of his errors had turned up on Trivial Pursuit cards.

Worth had always been worried that some other writer would boost his facts, so he had intentionally planted a fake "fact" in his text: that the title character in the TV series *Columbo* had the first name "Frank." (In truth, the show never revealed Peter Falk's character's first name.) When the Frank Columbo "fact" showed up in a Trivial Pursuit question, Worth began calculating just how much information the game's creators had borrowed from his book.

After determining that roughly 32 percent of Trivial Pursuit's questions had originally appeared in his book, Worth sued the game's creators and manufacturers for a cool $300

Irving Berlin donated the royalties from "God Bless America" to the Boy & Girl Scouts of America.

million in October 1984. The case made it all the way to the Supreme Court, but each time the result was the same. The creators of Trivial Pursuit admitted they had used Worth's book as a source, but they argued that no one can copyright facts. Unfortunately for Worth, the courts agreed each time, and he never got his $300 million payday.

Ralph Lauren Undresses Magazine

When *Polo* magazine launched in 1975, fashion designer Ralph Lauren didn't worry about the name being similar to his most famous clothing line, Polo. *Polo* was devoted to the actual sport of the same name, and it was even the official magazine of the United States Polo Association. Lauren felt that nobody would confuse a magazine about polo with his line of preppy clothing.

For 22 years, this arrangement worked out pretty well. Then in 1997 a media group bought *Polo* and relaunched it as a glossy mag devoted to fashion and high-end lifestyle pieces. That description sounds awfully familiar to anyone who's ever bought a Ralph Lauren shirt, even if they can't tell a polo pony from a donkey. Lauren sued for trademark infringement on the grounds that an unrelated fashion magazine sharing its name with his clothing line was confusing.

A federal judge agreed with Lauren, and in 1999 *Polo* received word that it had 90 days to stop publishing under that name. *Polo* wasn't quite dead, though. A 2001 appeal reversed the ruling, and the magazine made a comeback, complete with a disclaimer it wasn't affiliated with Polo Ralph Lauren.

In 2006, Merv Griffin estimated he'd made $70 million in royalties from the *Jeopardy!* theme song alone.

THE WORLD'S A TROPHY CASE

WHERE FIVE OSCAR WINNERS KEEP THE LITTLE GUY

1 In the fridge.

Timothy Hutton won a Best Supporting Actor Oscar for *Ordinary People* in 1981. He recently said he and his sister were having a party at his house a few years ago when she decided to stick his Oscar in the fridge so that people going to grab a beer would be entertained. He enjoyed the joke and has kept it there ever since.

2 At his dad's hardware store.

Patrons of the J. M. Stewart & Co. hardware store in Indiana, Penn., probably wouldn't have been that surprised to find an Academy Award hanging out among the saws and paintbrushes. When Jimmy Stewart won his Oscar in 1941 for his lead role in *The Philadelphia Story*, he promptly sent his award to his father, who displayed it in the family store for 25 years.

3 At home, dressed up in Barbie clothes.

John Lasseter of Pixar has two Oscars under his belt—Best Animated Short ("Tin Toy" in 1989) and a Special Achievement Award for *Toy Story* in 1996. "We discovered that Barbie clothes actually fit pretty well," he once commented. "Oscar's shoulders are a little broader, so we let them out a little." Guess that happens when you have five kids.

4 In the underwear drawer.

Because that's where you keep all valuables, isn't it? After Kevin Costner won Best Director for *Dances with Wolves* in 1991 (it also won Best Picture), he stashed his award in his underwear drawer for a few years. He has since had a trophy case built and keeps Oscar in there. Do you suppose he keeps his Razzie for *The Postman* in there as well?

5 On the dresser, used as a hat stand.

We suppose when you start racking up the Oscars like Jack Nicholson has, you have to find inventive uses for them. Rumor has it Jack uses one of his three to hold a chapeau or two.

Because of a metal shortage during World War II, Oscar statuettes were briefly made of painted plaster.

10 LISTS

FOR PEOPLE WHO CAN'T WRITE GOOD

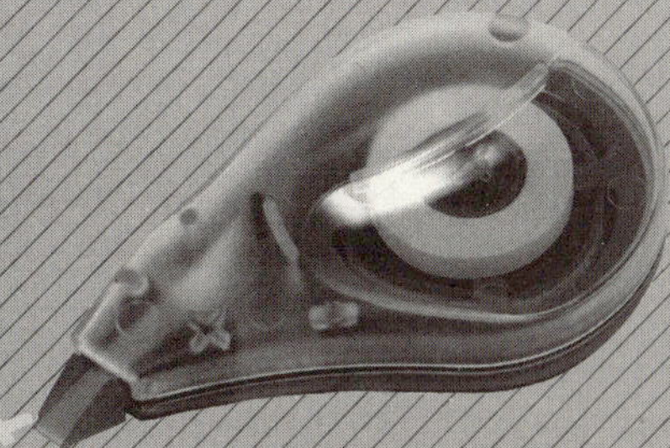

SIX WORKS OF LITERATURE THAT WERE REALLY HARD TO WRITE

1 The Story That Will Never Be an e-Book
Gadsby by Ernest Vincent Wright

Some might call *Gadsby* a "love" story. But Ernest Vincent Wright wouldn't have used that word. Instead, he described his novel as a story of "strong liking" and "throbbing palpitation." That's because in 1939, Wright gave himself one restriction: He promised to write *Gadsby* without using the letter E. Wright wanted to prove that a great author could work around such a restriction and still tell a gripping story. To prevent any stray Es from entering the text, he tied down his typewriter's E key, and then put his expansive vocabulary to the test.

The result is an astounding feat of verbal gymnastics. While vividly describing a wedding scene, Wright manages to avoid the words "bride," "ceremony," and even "wedding" (he calls it "a grand church ritual"). To explain away the verbosity of the language, he uses a narrator whose poor command of English and circumlocution even irritates the story's other characters.

When the book was announced, one skeptic attacked Wright in a letter, claiming that the feat was impossible. "All right," replied Wright in the book's intro, "the impossible has been accomplished." Sadly, Wright didn't live long enough to revel in *Gadsby*'s critical acclaim. He died the year the book was published.

At 1,536 pages in its abridged version, Samuel Richardson's 1748 *Clarissa, or, the History of a Young Lady* is considered the longest novel in the English language.

2 Six Powerful Words
"Baby Shoes" by Ernest Hemingway

According to legend, Ernest Hemingway created the shortest short story ever told. While having lunch at New York City's famous Algonquin Round Table, Hemingway bragged that he could write a captivating tale—complete with beginning, middle, and end—using only six words. His fellow writers refused to believe it, each betting $10 that he couldn't do it. Hemingway quickly scribbled six words down on a napkin and passed it around. As each writer read the napkin, they conceded he'd won. Those six words? "For sale. Baby shoes. Never worn."

While the anecdote may be apocryphal, whoever did write "Baby Shoes" has forced writers forever after to consider the economy of words. Today, the work has inspired countless six-word memoir and story competitions, proving that a story's brevity is no limit to its power.

3 James Joyce's Deaf Translation Jam
Finnegans Wake by James Joyce

James Joyce wrote his final novel, *Finnegans Wake*, during a 17-year period in Paris, finishing the work just two years before his death in 1941. During that time, Joyce was nearly blind, so he dictated his stream-of-consciousness prose to his friend, Samuel Beckett.

This process led to some unexpected results. For example, during one session, Joyce heard a knock at the door, which was too quiet for Beckett to perceive. Joyce yelled to the visitor, "Come in!" so Beckett added "Come in!" to the manuscript. When Beckett later read the passage back to Joyce, the author decided that he liked it better that way.

After several such sessions, *Finnegans Wake* became one of the most impenetrable works of English literature. But the experience didn't just affect Joyce's novel; it seemed to have a lasting effect on Beckett's writing, as well. Beckett would go on to become a leading playwright in the Theatre of the Absurd, where his characters often spent their entire time onstage sitting in the middle of nowhere, hoping that someone would hear their voice.

4 The Art of Writing by Committee
The President's Mystery Story by Franklin Roosevelt and Seven Other Novelists

Many American presidents have written books, but only Franklin Roosevelt has contributed to a mystery novel. At a White House dinner in 1935, Roosevelt pitched his story idea to author Fulton Oursler. Roosevelt's tale started like this: A man named Jim Blake is trapped in a stale marriage and a boring job. He dreams of running off with $5 million and starting over with a new identity.

Unfortunately, the President hadn't worked out one major plot point: How does a man with $5 million disappear without being traced?

Johnny Cash's "A Boy Named Sue" was penned by Shel Silverstein, the beloved children's book author who wrote *Where the Sidewalk Ends*.

To solve the problem, Oursler formed a committee of five other top mystery writers: Rupert Hughes, Samuel Hopkins Adams, Rita Weiman, S. S. Van Dine, and John Erskine. Each author wrote a chapter and ended it with Jim Blake in a terrible situation, which the next author was left to resolve. Despite being the work of a Washington committee, the end result was surprisingly successful. *The President's Mystery Story* was serialized in a magazine, published as a book, and even turned into a movie in 1936.

Yet the writers never came up with a solution to Roosevelt's original problem. That didn't happen until 1967, when Erle Stanley Gardner wrote a final chapter to a new edition of the book. In it, the secret to Jim Blake's mysterious disappearance is discovered by Gardner's most famous character, Perry Mason.

5 The Most Visionary Story Ever Told
Futility by Morgan Robertson

Perhaps the most meticulously prophetic work of literature is Morgan Robertson's short and poorly written novel, *Futility*. In it, Robertson describes the maiden voyage of a British luxury liner called the *Titan*, which claims to be unsinkable, but sinks anyway after hitting an iceberg. Nearly every detail resembles the story of the *Titanic*. Of course, nobody thought about that when *Futility* was released in 1898, a full 14 years before the *Titanic* set sail.

***Futility* wasn't Robertson's only prescient piece of literature. In 1912, three years before his death, he wrote *Beyond the Spectrum*. Robertson's story predicted a Japanese sneak attack on an American fleet in Hawaii, and the resulting war between the two countries.**

Writing by Ear
Anguish Languish by Howard L. Chace

Sinker sucker socks pants, apocryphal awry. If those words don't make sense together, try saying them out loud: "Sing a song of sixpence, a pocketful of rye." Now imagine a whole book written like this, and you've got Howard L. Chace's 1940 collection of nursery rhymes and fairy tales, *Anguish Languish*. The work contains classics such as Marry Hatter Ladle Limb and Ladle Rat Rotten Hut, which begins with the immortal line, "Wants pawn term, dare worsted ladle gull hoe lift wetter murder inner ladle cordage."

Although *Anguish Languish* is playful, there was also a serious side to it. As a French professor, Chace used the stories to illustrate that, in spoken English, intonation is almost as important to the meaning as the words themselves.

Honore de Balzac reportedly drank 50 cups of coffee a day to fuel his writing.

PANSIES AND STARKILLERS

WHAT 10 FICTIONAL CHARACTERS WERE ALMOST CALLED

1. Scarlett O'Hara was nearly a Pansy. The iconic character didn't receive her iconic name until just before *Gone with the Wind* went to print.

2. In early drafts of *Breakfast at Tiffany's*, Holly Golightly was named Connie Gustafson.

3. Bram Stoker's notes on *Dracula* reveal that he had been referring to his famous vampire as "Count Wampyr." During research, Stoker came across Vlad II of Wallachia, who went by the name Vlad Dracul. The story so intrigued Stoker that he changed his character's name.

German poet Friedrich Schiller always kept a bowl of rotting apples on his desk. He said the smell helped him write.

4. Similarly, Sir Arthur Conan Doyle made notes that indicated he had been considering the name "Sherringford" for Sherlock Holmes.

5. If that doesn't throw you for enough of a loop, consider this: Holmes' assistant was originally named "Ormond Sacker." Arthur Conan Doyle decided the name was a bit too bizarre and changed it to the decidedly duller "John H. Watson."

6. **Batman's alter ego, Bruce Wayne, borrows his surname from Revolutionary War leader General "Mad" Anthony Wayne. The hero's creators were looking for sturdy, historical names that suggested gentry and entitlement. Before settling on "Bruce Wayne," they also considered Bruce Adams and Bruce Hancock. "Bruce," by the way, came from Scottish patriot Robert Bruce.**

7. If she hadn't been Nancy Drew, the plucky young heroine could have been Stella Strong, Diana Drew, Diana Dare, Nan Nelson, Helen Hale, or Nan Drew.

8. Small Sam, Little Larry, and Puny Pete were all in the running before Charles Dickens settled on "Tiny Tim" for the sickly sad sack in *A Christmas Carol*.

9. Little Orphan Annie was nearly Little Orphan Otto until Harold Gray's publisher suggested his character looked more female than male and told him to stick a skirt on it.

10. It may have been a much different story if George Lucas had gone with his original "Luke Starkiller" name. Although the Skywalker name prevailed, "Starkiller" has since popped up for other characters.

ntence that uses every letter in the alphabet. One famous pangram? "The quick over the lazy dog."

RENAMES OF THE DAY

WHAT EIGHT CLASSIC BOOKS WERE ALMOST CALLED

Settling on a title can be tough. We almost called this book Think Yourself Thin!

❶ ***The Great Gatsby***

F. Scott Fitzgerald considered quite a few titles, including *Trimalchio in West Egg*; *Among Ash-Heaps and Millionaires*; *On the Road to West Egg*; *Under the Red, White, and Blue*; *Gold-Hatted Gatsby*; and *The High-Bouncing Lover*.

❷ ***1984***

George Orwell's publisher didn't feel the title *The Last Man in Europe* was terribly commercial and recommended using the other title he had been kicking around—*1984*.

A word that can be its own antonym is called a contronym. For example, "cleave" can mean to sever or to cling.

3 *Atlas Shrugged*

Before it was *Atlas Shrugged*, it was *The Strike*, which is how Ayn Rand referred to her magnum opus for quite some time. In 1956, a year before the book was released, she decided the title gave away too much plot detail. Her husband suggested *Atlas Shrugged*, and the title stuck.

4 *Dracula*

The title of Bram Stoker's famous Gothic novel sounded more like a spoof before he landed on *Dracula*—one of the names Stoker considered was *The Dead Un-Dead*.

5 *Catch-22*

Thank Frank Sinatra for the phrase "Catch-22." Sort of, at least. Author Joseph Heller originally called his novel *Catch-11*, but he scrapped that title because the original *Ocean's Eleven* movie was still in theaters. Heller also considered *Catch-18*, but, again, a recent publication made him switch titles to avoid confusion: Leon Uris' *Mila 18*. Heller finally settled on the number 22. It was finally chosen because of its repetition of "2."

6 *Dubliners*

James Joyce originally called his book of short stories *Ulysses in Dublin*. Guess he decided that that title was so loaded that he could get two whole books out of it.

7 *To Kill a Mockingbird*

To Kill a Mockingbird was simply *Atticus* before Harper Lee decided the title focused too narrowly on one character.

8 *Pride and Prejudice*

Jane Austen's novel probably would have done just as well with its original title: *First Impressions*.

That thing you use to dot your lowercase "i" is called a tittle.

COGITO ERGO SOMETHING

EIGHT LATIN PHRASES YOU PRETEND TO UNDERSTAND

Whether you're deciphering a cryptic state seal or trying to impress your Catholic in-laws, knowing some Latin has its advantages. But the operative word here is "some." We'll start you off with eight phrases that have survived the hatchet men of time (in all their pretentious glory).

Caveat Emptor

(KAV-ee-OT emp-TOR): "Let the buyer beware"

Before money-back guarantees and 20-year warranties, *caveat emptor* was indispensable advice for the consumer. These days, it'd be more fitting to have it tattooed on the foreheads of used-car salesmen, infomercial actors, and prostitutes. For extra credit points, remember that caveat often makes solo appearances at cocktail parties as a fancy term for a warning or caution. Oh, and just so you know, *caveat lector* means "let the reader beware."

The Latin version of *Winnie the Pooh* (*Winnie ille Pu*) made the *New York Times* bestseller list in 1960.

Persona Non Grata

(puhr-SOH-nah non GRAH-tah): "An unacceptable person"

Remember your old college buddy, the one everybody called Chugger? Now picture him at a debutante ball, and you'll start to get a sense of someone with persona non grata status. The term is most commonly used in diplomatic circles to indicate that a person is unwelcome due to ideological differences or a breach of trust. Sometimes, the tag refers to a pariah, a ne'er-do-well, a killjoy, or an interloper, but it's always subjective. Back in 2004, Michael Moore was treated as a persona non grata at the Republican National Convention. Bill O'Reilly would experience the same at Burning Man.

Habeas Corpus

(HAY-bee-as KOR-pus): "You have the body"

When you wake up in the New Orleans Parish Prison after a foggy night at Mardi Gras, remember this one. In a nutshell, habeas corpus is what separates us from savages. It's the legal principle that guarantees an inmate the right to appear before a judge in court, so it can be determined whether or not that person is being lawfully imprisoned. It's also one of the cornerstones of the American and British legal systems. Without it, tyrannical and unjust imprisonments would be possible. In situations where national security is at risk, however, habeas corpus can be suspended.

4

Cogito Ergo Sum

(CO-gee-toe ER-go SOME): "I think, therefore I am"

When all those spirited mental wrestling matches you have about existentialism start growing old (yeah, right!), you can always put an end to the debate with cogito ergo sum. René Descartes, the 17th-century French philosopher, coined the phrase as a means of justifying reality. According to him, nothing in life could be proven except one's thoughts. Well, so he thought, anyway.

5

E Pluribus Unum

(EE PLUR-uh-buhs OOH-nuhm): "Out of many, one"

Less unique than it sounds, America's original national motto, *e pluribus unum*, was plagiarized from an ancient recipe for salad dressing. In the 18th century, haughty intellectuals were fond of this phrase. It was the kind of thing gentlemen's magazines would use to describe their year-end editions. But the term made its first appearance in Virgil's poem "Moretum" to describe salad dressing. The ingredients, he wrote, would surrender their individual aesthetic when mixed with others to form one unique, homogenous, harmonious, and tasty concoction. As a slogan, it really nailed that whole cultural melting pot thing we were going for. And while it continues to appear on U.S. coins, "In God We Trust" came along later (officially in 1956) to share the motto spotlight.

Finnish academic Dr. Jukka Ammondt has developed a following by singing Elvis' songs in Latin. Want to hear "Love Me Tender"? Ask for "*Tenere me ama.*"

Quid Pro Quo

(kwid proh KWOH): "You scratch my back, I'll scratch yours"

Given that quid pro quo refers to a deal or trade, it's no wonder the Brits nicknamed their almighty pound the "quid." And if you give someone some quid, you're going to expect some quo. The phrase often lives in the courtroom, where guilt and innocence are the currency. It's the oil that lubricates our legal system. Something of a quantified value is traded for something of equal value; elements are parted and parceled off until quid pro quo is achieved.

Ad Hominem

(ad HAH-mi-nem): "To attack the man"

In the world of public discourse, ad hominem is a means of attacking one's rhetorical opponent by questioning his or her reputation or expertise rather than sticking to the issue at hand. Translation: Politicians are really good at it. People who resort to ad hominem techniques are usually derided as having a diluted argument or lack of discipline. If pressed, they'll brandish it like a saber and refuse to get back to the heart of the matter. Who said the debate team doesn't have sex appeal?

Sui Generis

(SOO-ee JEN-er-is): "Of its own genus," or "Unique and unable to classify"

Frank Zappa, the VW Beetle, cheese in a can: Sui generis refers to something that's so new, so bizarre, or so rare that it defies categorization. Granted, labeling something "sui generis" is really just classifying the unclassifiable. But let's not overthink it. Use it at a dinner party to describe Andy Kaufman, and you impress your friends. Use it too often, and you just sound pretentious.

Poet Heinrich Heine left his wife his fortune with one catch: she had to remarry. "Then there will be at least one man to regret my death."

DR. SEUSS TAUNTS THE NERDS

SIX WORDS INVENTED BY AUTHORS

1. **"Nerd"** may be a common insult (or term of endearment, depending on your tone) these days, but before Dr. Seuss published it in 1950's *If I Ran the Zoo*, people had to make do with "square" and "drip" instead. At least, they did according to *Newsweek*, which ran an article in 1951 defining the new slang term.

2. **"Bump"** first appeared in Shakespeare's *Romeo and Juliet*. Shakespeare coined a ridiculous number of words, actually, although some historians and linguists think certain words just get attributed to Shakespeare even if he didn't really invent them.

3. It may be hard to believe, but the word **"quark"** first appeared in James Joyce's *Finnegans Wake*. Scientist Murray Gell-Mann had been thinking about calling the unit "kwork," but when he found the invented word in the Joyce classic, he knew he had discovered the spelling he wanted to use.

4. If you have a **Tween** in your life, you can thank J.R.R. Tolkien that you have something to define them with. In *The Fellowship of the Ring*, Tolkien claimed a Tween was a Hobbit between the ages of 20 and 33 (33 being when Hobbits come of age). There's some debate as to whether the word existed prior to this reference or not, however—the *Oxford English Dictionary* does not give him credit.

5. **"Runcible spoon"** was created for *The Owl and the Pussycat* by author Edward Lear. He had no particular meaning for the word runcible—Lear also referred to "a runcible hat," a "runcible cat," a "runcible goose," and a "runcible wall." But since it has entered somewhat common vernacular, "runcible spoon" sometimes refers to a grapefruit spoon, a spork, or a sort of flattened ladle (which is what it looks like in the illustration that accompanies the poem).

6. Enjoy goofing around in **cyberspace**? Thank sci-fi writer William Gibson. He didn't invent cyberspace itself, but the word first appeared in his 1982 short story "Burning Chrome."

The string of typographical symbols comic strips use to indicate profanity ("$%@!") is called a grawlix.

GREAT UNEXPECTATIONS

THE LITTLE-KNOWN STORIES OF FIVE FAMOUS AUTHORS

Charles Dickens Gets a Grip

A number of pets graced the Dickens household over the years, including all manner of dogs, cats, and ponies. But Charles' favorite pets were his two ravens, both known as Grip. Dickens was particularly devoted to Grip I, going so far as to write the bird into his 1841 mystery novel, *Barnaby Rudge*. This same talkative bird reportedly was the inspiration for Edgar Allan Poe's famous poem, "The Raven," published four years later. Upon Grip I's demise, Dickens had his beloved bird stuffed. These days, Grip can be seen at the Free Library of Philadelphia's Rare Books Department, where he stands guard over the Poe and Dickens collections.

2 Langston Hughes: The Busboy Poet

Poet, playwright, novelist, essayist, and all-around literary luminary, Langston Hughes achieved fame during the Harlem Renaissance. But before that, Hughes was a struggling young writer, working menial jobs to support his burgeoning poetry habit. In 1925, while working at a restaurant in Washington, D.C., Hughes tucked a few of his poems under the dinner plate of then-reigning poet Vachel Lindsay. Lindsay shared the poems during his reading that night, and in the morning, Hughes was crowned Lindsay's new discovery, the "busboy poet." Hughes went on to become one of America's most prolific authors.

Thriller writer James Patterson was formerly an adman. He coined the slogan "I'm a Toys 'R' Us kid."

3 Thomas Hardy Puts His Heart into It

When British poet and novelist Thomas Hardy died on January 11, 1928, his literary contemporaries decided he was too important to be buried in his hometown's simple churchyard. But the good people of Dorset, where Hardy had spent nearly all of his 88 years, vehemently disagreed. So the two groups reached a grisly compromise. The author's body was cremated, and his ashes were interred in the Poet's Corner at Westminster Abbey. Hardy's heart, on the other hand, was placed inside a small casket and buried beside the grave of his first wife in a Dorset churchyard. To this day, a rumor persists that the author's heart was accidentally devoured by his housekeeper's cat, and that the heart of a pig was buried in its place.

4 Edith Wharton, War Hero

Edith Wharton, winner of the Pulitzer Prize for her 1920 novel *The Age of Innocence*, is famous for her vivid stories and novels about upperclass society in the late 19th century. It was a setting she knew well, coming from a wealthy and distinguished New England family. But the high society author had a lesser-known career as a humanitarian. During World War I, Wharton traveled to the Western Front in France, both to write about the battlefields for American publications and to help the Red Cross create hostels and schools for those displaced by war. In 1916, she was awarded the Legion of Honor, France's highest civilian medal, years before the height of her literary career.

5 Horatio Alger, Jr.: Chased Out of Town

Apparently, the author of more than 120 "rags-to-riches" books featuring hardworking, highly moral young heroes was also an admitted pederast. Before finding success as an author, Alger was a minister at a Unitarian Church in Brewster, Mass., where he was accused of sexually assaulting two young boys. Alger admitted his guilt, but left town before the news hit the street. Later, he wound up in New York City, where he penned hundreds of bestselling books for and about young boys, which went on to grace the shelves of homes, schools, and church libraries across America.

Poet Robert Frost dropped out of both Harvard and Dartmouth as an undergrad.

PUTTING A GOOD PHRASE ON IT

FIVE FAMOUS PHRASES PEOPLE OWN

1. **"Three-Peat"**

Coach Pat Riley registered this one when his Los Angeles Lakers began their campaign for a third straight NBA title in 1988. While Riley's Showtime Lakers fell short of its three-peat, the coach raked in cash after the Chicago Bulls nabbed their third straight title in 1993 and used the phrase in their celebratory promotions.

2. **"Let's Get Ready to Rumble!"**

Boxing and wrestling announcer Michael Buffer owns his trademark call to arms. Some of his better brand extensions include a welcome message in New York cabs that encouraged riders to buckle up ("Let's get ready to rumble . . . for SAFETY!") and Kraft cheese crumbles ("Lets get ready to crumble!").

3. **"That's Hot"**

Paris Hilton has owned four different trademarks for her catchphrase that covered her right to use "That's Hot" on everything from clothing to alcoholic beverages. The latter came in hand when she used the phrase in promotions for a canned Italian sparkling wine called "Rich Prosecco."

4. **"Bam!"**

Celebrity chef Emeril Lagasse owned the rights to "Bam!"—exclamation point and all—for use on pretty much any item you might find in a kitchen. In 2008 he sold the phrase to Martha Stewart's company.

5. **"They Are Who We Thought They Were"**

When normally reserved former Arizona Cardinals coach Dennis Green's squad choked away a big game to the Chicago Bears in 2006, Green threw a monumental postgame tantrum. He also picked up a catchphrase. Green later trademarked his tirade's refrain—"They are who we thought they were"—for use on sports merchandise. Green has since abandoned his rights to the phrase.

The so-called *Wicked Bible* published in 1631 read, "Thou shalt commit adultery." Whoops!

SHIVER OUR TIMBERS!

THE NAUTICAL ROOTS OF SIX COMMON PHRASES

1 Three Sheets to the Wind

Many people are surprised to learn that this expression for drunkenness was born on the high seas. "Sheet" is the nautical term for the rope that controls the tension on a sail. If the sheets are loose on a three-masted ship, then the sails will flap uselessly in the wind, and the ship will drift out of control until the situation is corrected. Thus, the modern phrase "three sheets to the wind" has come to signify a person who is intoxicated to the point of being out of control.

Filibuster

The roots of the term "filibuster" can be traced to the pirates who prowled the shipping trade routes in the 17th, 18th, and 19th centuries. The Dutch word for pirate was *vrijbuiter*—a word that eventually led to the French term *flibustier* and the Spanish term *filibustero*. The British, however, pronounced it filibuster.

The longest filibuster in Senate history lasted 57 days. It was performed by a bloc of Southern senators to thwart the Civil Rights Act of 1964.

So how did the word for pirate become associated with obstructionist political tactics? It's still a bit of a mystery, but some historians speculate that, since pirates were an incessant, obstructing nuisance, they effectively blocked trade in many areas, just as politicians try to block legislation today.

❸ Slush Fund

Most people think this term originated in the smoke-filled boardrooms of corporate America. Surprisingly, however, it can be traced back to some clever ship cooks who saved the slushy mix of fat and grease that was left over after every meal. The slush would be stowed away in a secret hiding place until the ship returned to port. The cooks would then sell the fat to candle makers and other merchants, earning themselves a tidy sum in the process. Thus, the term "slush fund" refers to an illicit cash reserve.

❹ Under the Weather

Keeping watch on board sailing ships was a boring and tedious job, but the worst watch station was on the "weather" (windward) side of the bow. The sailor who was assigned to this station was subject to the constant pitching and rolling of the ship. By the end of his watch, he would be soaked from the waves crashing over the bow. A sailor who was assigned to this unpleasant duty was said to be "under the weather." Sometimes, these men fell ill and died as a result of the assignment, which is why today "under the weather" is used to refer to someone suffering from an illness. A related theory claims that ill sailors were sent below deck (or "under the weather") if they were feeling sick.

❺ Chew the Fat

Before refrigeration, salted beef and pork were staple foods aboard sailing vessels because they could be stored for long periods without spoiling. However, they were also tough and extremely difficult to eat. It often took a great deal of chewing just to soften up the meat and make it edible, which took a lot of time. So, in the spirit of multitasking, men would gather to discuss the day's events while they chewed their fatty, salt-cured meat. According to this theory, whenever people get together to gossip or chat, we say that they are "chewing the fat."

❻ Clean Bill of Health

The "Age of Sail" in the 18th and early 19th centuries was a glorious time in naval history marked by many epic battles on the high seas, but it was also a time of widespread disease. In order to receive permission to dock at a foreign port, ships were often required to show a bill of health—a document that stated the medical condition of their previous port of call, as well as that of everyone aboard. A "clean bill of health" certified that the crew and their previous port were free from the plague, cholera, and other epidemics. Today, a person with a "clean bill of health" has passed a doctor's physical or other medical examination.

Due to a shortage of raw materials like paper and leather—and an increase in wartime piety—the United States faced a Bible shortage in 1943.

NINE THINGS

MARK TWAIN DIDN'T SAY (AND NINE HE DID)

"The secret of getting ahead is getting started."

He DID say: "Never put off till tomorrow what may be done day after tomorrow just as well."

"It is better to keep your mouth shut and appear stupid than to open it and remove all doubt."

He DID say: "[He] was endowed with a stupidity which by the least little stretch would go around the globe four times and tie."

When Ulysses S. Grant was short on cash after leaving the White House, his buddy Twain convinced him to pen his memoirs to drum up some dough.

"Censorship is telling a man he can't have a steak just because a baby can't chew it."

He DID say: "When a Library expels a book of mine and leaves an unexpurgated Bible lying around where unprotected youth and age can get hold of it, the deep unconscious irony of it delights me and doesn't anger me."

4 **"Everybody talks about the weather, but nobody does anything about it."**

He DID say: "I reverently believe that the Maker who made us all makes everything in New England but the weather. I don't know who makes that, but I think it must be raw apprentices in the weather-clerk's factory who experiment and learn how, in New England, for board and clothes, and then are promoted to make weather for countries that require a good article, and will take their custom elsewhere if they don't get it."

5 **"I would have written a shorter letter, but I did not have the time."**

This was actually written by Blaise Pascal.
He DID say: "We write frankly and fearlessly but then we 'modify' before we print."

6 **"The coldest winter I ever spent was a summer in San Francisco."**

He DID say: "Cold! If the thermometer had been an inch longer we'd all have frozen to death."

7 **"There are three kinds of lies: lies, damned lies, and statistics."**

Twain himself denied inventing this quote and claimed Benjamin Disraeli was the one who created it. Twain did popularize the saying in the States, though.
He DID say: "Yes, even I am dishonest. Not in many ways, but in some. Forty-one, I think it is."

8 **"Be careful about reading health books. You may die of a misprint."**

He DID say: "A successful book is not made of what is in it, but of what is left out of it."

9 **"Twenty years from now you will be more disappointed by the things you didn't do than by the ones you did do."**

He DID say: "One cannot have everything the way he would like it. A man has no business to be depressed by a disappointment, anyway; he ought to make up his mind to get even."

Twain on his being born and dying in years when Halley's Comet passed: "The Almighty has said, no doubt: 'Now here are these two unaccountable freaks; they came in together, they must go out together.'"

QUEUE 'EM UP

10 "Q" WORDS THAT AREN'T "Q-U" WORDS

Q

You may think the letter "Q" can't go anywhere without his partner "U," but their relationship isn't completely monogamous.

1. **Bathqol**—Also sometimes spelled "bath quol," this one's a divine revelation given to certain Jewish teachers.

2. **Coq**—cock feathers on a woman's hat.

3. **Qiana**—This one's tricky. It used to be the trademarked name for a silky nylon that DuPont invented in the '60s, but like "Frisbee" it became so widely used that it's now a common noun.

4. **Qanun**—a type of harp.

5. **Qawwal**—a person who plays South Asian Qawwali music.

6. **Qintar or qindar**—a type of Albanian money.

7. **Qiviut**—the wool of a musk ox.

8. **Qwerty**—the keyboard layout that includes that sequence of letters.

9. **Umiaq**—a type of open boat.

10. **Qat**—a type of flowering plant.

One more for good measure: a niqāb is a veil worn by some Muslim women.

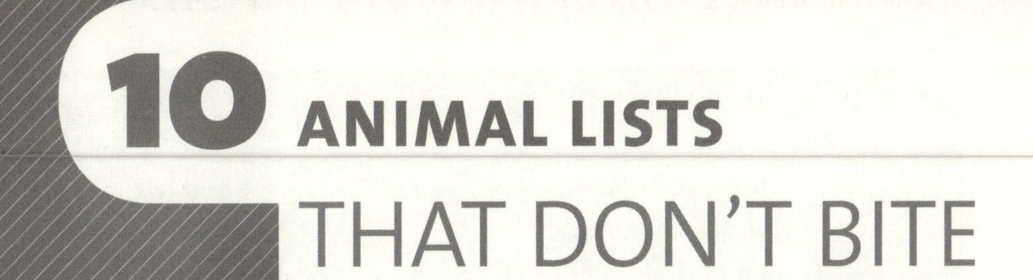

10 ANIMAL LISTS THAT DON'T BITE

DRILL SEEKERS

SEVEN CLEVER CRITTERS THAT ARE GREAT WITH TOOLS

Elephants Drink Bottled Water

Not only do elephants use branches to swat flies and scratch their backs, but they also use tools to plan for the future. In South Africa, biologist Hezy Shoshani observed a pachyderm chewing bark into a large ball and then using the ball to plug up a nearby watering hole. The result was an elephant-size water bottle! Later, the animal came back to the spot, removed the ball, and quenched his thirst again.

Dolphins Cover Their Mouths

In addition to bouncing balls on their noses, dolphins are also handy with sponges. Georgetown University researcher Janet Mann reported that bottlenose dolphins in Australia's Shark Bay have been seen carrying sea sponges in their mouths while fishing along the ocean floor. When they dig into the sand to stir up hidden fish, the sponges apparently act as a kind of mask. But, of the thousands of bottlenose dolphins identified in Shark Bay, only 41 have been observed doing this. Almost all of them were female, and the behavior seems to be something mothers teach their daughters.

Walruses' mustaches are full of nerve endings. They use their whiskers to find tasty shellfish on dark ocean floors.

Crows Have a Lot to Crow About

New Caledonian crows are widely renowned as the tool-using champs of the bird kingdom. To hunt for insects, they shape sticks into hooks and spears that allow them to probe tree crevices. They also modify those sticks into the correct size and shape by whittling them with a complex process of snips and tears. What's more, New Caledonian crows can make new tools out of old ones and pass along their new inventions to others. The only other creatures on Earth known to do this are humans.

4 Owls Make the Most Out of Cow Poop

Some burrowing owls have a strange habit of scattering cow manure in the ground around the entrances to their homes. Until recently, scientists thought this behavior evolved as a way to mask the owls' scent from potential predators. But researchers have determined that the cow manure actually functions as bait to lure dung beetles, one of the owls' favorite foods.

5 Vultures Cast Stones

Egyptian vultures love the taste of ostrich eggs, but they can't break the thick shells by just pecking at them. So hungry vultures go in search of rocks for the job, sometimes venturing up to 50 yards away. When they return, they dip their heads violently and hurl their rock at the egg, smashing open the shell. Surprisingly, this technique appears to be an innate behavior. When presented with tasty eggs, even vultures raised alone in captivity will go hunting for stones.

Chimps Build Nutcrackers

Chimpanzees of the Ivory Coast's Tai Forest are the Bob Vilas of their species. In order to crack open the hard oil-palm nuts they adore, the chimps use two tools at once. First, they place a nut on a flat stone for traction, then they smash it with a pointed hammer-like stone. The skill takes young chimps several years to master, but once they get the hang of it, they'll store their favorite tool sets in certain places and bring their nuts there for cracking. A recent archaeological dig found that Tai Forest chimps have been making nutcrackers like these for 4,000 years.

Herons Go Fishing

Like Jane Goodall's chimpanzees, wild green-backed herons "fish" for their food. Using insects, feathers, or even flowers, they drop their clever bait into the water and then gobble up the curious fish that come to the surface for a meal. Herons can be remarkably persistent fishermen, too. Reportedly, one researcher in Africa watched a heron drop the same bait into the water 28 times in a row before a fish finally bit.

Switzerland was the last country to stop using carrier pigeons. The Swiss kept 30,000 of the birds around until the mid-1990s.

PET SCANS

THREE SUPER-ANIMALS KEEPING AN EYE ON TERROR

1 Bluegill Fish Test the Waters

During a meeting in 1960, Chicago Mayor Richard J. Daley presented Japanese Emperor Akihito with bluegill fish. It was a gift the Japanese people would later wish they never received. Since being introduced into Japanese fisheries, the bluegill has bred rapidly and killed native species. Even worse, the Japanese generally find its taste repugnant.

While it's marginally more popular in the United States, it's still not foodies' freshwater fish of choice. It is the fish of choice for the U.S. Army, though. Researchers created IAC 1090 Intelligent Aquatic Biomonitoring System, which uses eight juvenile bluegills to detect changes in the water. Each fish resides in its own stall about the width of a mail slot, and sensors monitor the fishes' breathing. If six of the eight bluegill show abnormalities, technicians know something is up.

Officials swap out bluegills—perfectly matched for the task because they are sedentary and sensitive to contaminants—so the fish aren't affected by their duties for long periods of time. New York, Washington, D.C., and San Francisco have used the fish to protect their water sources; the fish in New York alerted officials to a diesel leak before it became calamitous.

2 Release the Wasp Hounds!

Training dogs to sniff out drugs, bombs, or chemicals takes months and costs thousands of dollars. But University of Georgia researcher Glen Rains and the U.S. Department of Agriculture's Joe Lewis discovered that wasps could easily replace dogs' sniffing abilities. The researchers created a handheld device called the "Wasp Hound," which is a 15-inch cylinder with a vent at one end. When the parasitic *Microplitis croceipes* wasps smell their target they congregate at the vent, alerting humans. Rains and Lewis claim the wasps can also sniff out food toxins, crop fungus, bodies, drugs, and even 2,4-DNT, a volatile component in dynamite that dogs struggle to smell.

A team of wasps only costs $100 and takes as little as five minutes to train. Using basic Pavlovian methods, researchers associate the target smell with food so that the wasps are sufficiently motivated. It's a short career, though; the wasps only work for 48 hours before researchers release them.

It's estimated that 95 percent of the world's lab mice are descended from mice born in the Jackson Laboratory in Bar Harbor, Maine.

3 Crime-Fighting Bees

When Nigel Raine decided to study how serial killers track their victims, he placed tiny RFID chips on honeybees. Raine hypothesized that serial killers forage for victims much like predatory animals, such as sharks, or pollinators, such as bees. The RFID chips showed that bees pollinate plants near their hive, but not too close. The insects created a buffer zone between the hive and their feeding grounds, which protected the hive from predators and parasites. Similarly, serial killers feel comfortable preying in their neighborhoods, but not too close to their homes.

How will the police use this to find their serial killers? By creating a model of hunting, which criminologists can use to understand how serial killers work and create geographical profiles. With any luck, the models will help police track serial killers from crime scenes back to their homes.

And One Super-Animal That's Helping Ease the Pain

Unfortunately for leech salesmen, bloodletting has fallen out of fashion in recent centuries. Business could start booming again soon, though. Researchers from Germany's Essen-Mitte Clinic have discovered that leeches can soothe achy joints. Slap four leeches on your knee, and after 80 minutes the pain and stiffness of osteoarthritis melts away. Of the 16 patients in a clinical trial, the 10 who received leech therapy felt instant relief after application, and the comfort lasted for four weeks, leading the scientists to believe that the leeches' saliva works as an anti-inflammatory.

Sweden did not air the kangaroo-centric TV show *Skippy*, saying the series gave "a misleading impression of an animal's ability."

AMERICA'S GOT TALONS!

FIVE TECHNOLOGIES LIFTED FROM THE ANIMAL KINGDOM

1 Holy Bat Cane!

It sounds like the beginning of a bad joke: A brain expert, a bat biologist, and an engineer walk into a cafeteria. But that's exactly what happened when a casual meeting of the minds at England's Leeds University led to the invention of the Ultracane, a walking stick for the blind that vibrates as it approaches objects.

The cane works using echolocation, the same sensory system that bats use to map out their environments. It gives off 60,000 ultrasonic pulses per second and then listens for them to bounce back. When some return faster than others, that indicates a nearby object, which causes the cane's handle to vibrate. Using this technique, the cane not only "sees" objects on the ground, such as trash cans and fire hydrants, but also senses things above, such as low-hanging signs and tree branches. And because the cane's output and feedback are silent, people using it can still hear everything going on around them.

2 Puff the Magic Sea Sponge

The orange puffball sponge isn't much to look at; it's basically a Nerf ball resting on the ocean floor. It has no appendages, no organs, no digestive system, and no circulatory system. It just sits all day, filtering water. And yet, this unassuming creature might be the catalyst for the next technological revolution.

The "skeleton" of the puffball sponge is a series of calcium and silicon lattices. Actually, it's similar to the material we use to make solar panels, microchips, and batteries—except that when humans make them, we use tons of energy and all manner of toxic chemicals. Sponges do it better. They simply release special enzymes into the water that pull out the calcium and silicon and then arrange the chemicals into precise shapes.

Daniel Morse, a professor of biotechnology at the University of California, Santa Barbara, studied the sponge's enzyme technique and successfully copied it in 2006. He's already made a number of electrodes using clean, efficient sponge technology. And now, several companies are forming a multimillion-dollar alliance to commercialize similar products. In a few years, when solar panels are suddenly on every rooftop in America and microchips are sold for a pittance, don't forget to thank the little orange puffballs that started it all.

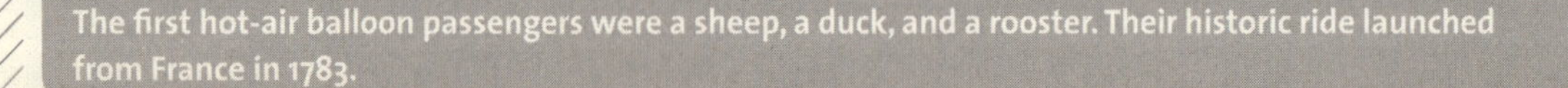

3 The Secret Power of Flippers

One scientist thinks he's found part of the solution to our energy crisis deep in the ocean. Frank Fish, a fluid dynamics expert and marine biologist at Pennsylvania's West Chester University, noticed something that seemed impossible about the flippers of humpback whales. Humpbacks have softball-size bumps on the forward edge of their limbs, which cut through the water and allow whales to glide through the ocean with great ease. But according to the rules of hydrodynamics, these bumps should put drag on the flippers, ruining the way they work.

Professor Fish decided to investigate. He put a 12-foot model of a flipper in a wind tunnel and witnessed it defy our understanding of physics.The bumps, called tubercles, made the flipper even more aerodynamic. It turns out that they were positioned in such a way that they actually broke the air passing over the flipper into pieces, like the bristles of a brush running through hair. Fish's discovery, now called the "tubercle effect," not only applies to fins and flippers in the water, but also to wings and fan blades in the air.

Based on his research, Fish designed bumpy-edge blades for fans, which cut through air about 20 percent more efficiently than standard ones. He launched a company called Whalepower to manufacture them and will soon begin licensing its energy-efficient technology to improve fans in industrial plants and office buildings around the world. But Fish's big fish is wind energy. He believes that adding just a few bumps to the blades of wind turbines will revolutionize the industry, making wind more valuable than ever.

4 Consider the Lobster Eye

There's a reason X-ray machines are large and clunky. Unlike visible light, X-rays don't like to bend, so they're difficult to manipulate. The only way we can scan bags at airports and people at the doctor's office is by bombarding the subjects with a torrent of radiation all at once—which requires a huge device.

But lobsters, living in murky water 300 feet below the surface of the ocean, have "X-ray vision" far better than any of our machines. Unlike the human eye, which views refracted images that have to be interpreted by the brain, lobsters see direct reflections that can be focused to a single point, where they are gathered together to form an image. Scientists have figured out how to copy this trick to make new X-ray machines. The Lobster Eye X-ray Imaging Device (LEXID) is a handheld "flashlight" that can see through three-inch-thick steel walls.

The device shoots a small stream of low-power X-rays through an object, and a few come bouncing back off whatever is on the other side. Just as in the lobster eye, the returning signals are funneled through tiny tubes to create an image. The Department of Homeland Security has already invested $1 million in LEXID designs, which it hopes will be useful in finding contraband.

In colonial America, lobster wasn't exactly a delicacy. In fact, it was so cheap and plentiful it was often served to prisoners.

Picking Up the Bill

The bill of the toucan is so large and thick that it should weigh the bird down. But as any Froot Loops aficionado can tell you, Toucan Sam gets around. That's because his bill is a marvel of engineering. It's hard enough to chew through the toughest fruit shells and sturdy enough to be a weapon against other birds, and yet, the toucan bill is only as dense as a Styrofoam cup.

Marc Meyers, a professor of engineering at the University of California at San Diego, has started to understand how the bill can be so light. At first glance, it appears to be foam surrounded by a hard shell, kind of like a bike helmet. But Meyers discovered that the foam is actually a complicated network of tiny scaffolds and thin membranes. The scaffolds themselves are made of heavy bone, but they are spaced apart in such a way that the entire bill is only one-tenth the density of water. Meyers thinks that by copying the toucan bill, we can create car panels that are stronger, lighter, and safer. Toucan Sam was right: Today we're all following his nose.

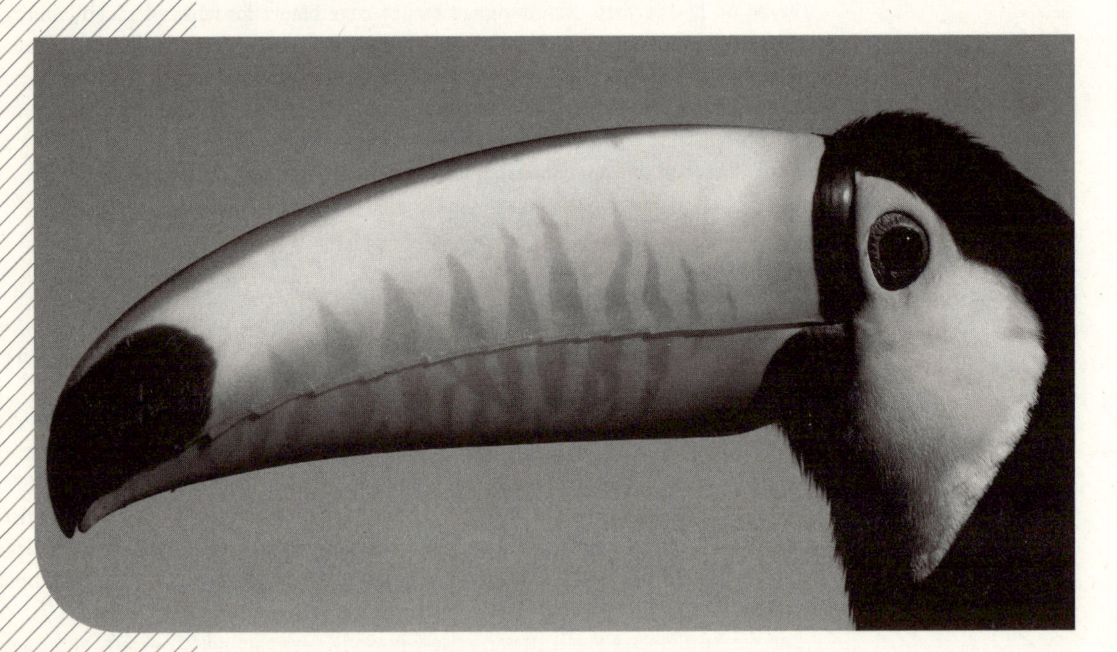

Redondo Beach, California, adopted the Goodyear Blimp as the city's official bird in 1983.

SEE SPOT RUN

THREE INCREDIBLE EXPLANATIONS ON HOW YOUR DOG WORKS

We can't tell you how to housebreak your dog, but we can answer some other lingering questions.

1 Why Do My Dog's Feet Smell Like Corn Chips or Popcorn?

Don't worry! "Frito feet," as the fragrant paws are sometimes known, are perfectly natural, and the explanation has nothing to do with stomping around in corn chips. First, consider this: The skin of most animals is home to a lot of microbes. Even a relatively clean human has around a hundred thousand bacteria on every square centimeter of his skin. Like we said, a *lot* of bacteria.

Dogs' feet are a great place for bacteria and yeast to take up residence because there's a lot of moisture and little to no air circulation in the folds and pockets of skin between toes and foot pads. In fact, bacteria flock there and reproduce with exuberance.

All these microorganisms give off their own distinct odors. The popcorn/corn chip smell on some dogs could be the fault of yeast or the bacterium *Proteus*, which are both known for their sweet, corn tortilla smell. Or it could be the bacterium *Pseudomonas*, which smells a little fruitier, but pretty close to popcorn to most noses.

These little critters are an integral part of life on Earth, and there are worse things the bacteria could be doing besides making your dog smell like delicious snacks.

2 Is My Dog's Mouth Really Cleaner Than My Own?

Comparing a dog's mouth with a human's mouth is sort of like comparing apples and oranges—really filthy apples and really filthy oranges. Both species' mouths are hot, damp places teeming with roughly equal populations of bacteria. Neither would be described as clean, and any question of comparative cleanliness is irrelevant because so much of that bacteria is species-specific. Most of the germs in your dog's mouth aren't going to be a problem during a big, wet doggie kiss. You're more likely to run into trouble kissing another human than you are a dog, because bacteria from a person's mouth will feel equally at home in yours.

Dalmatians take their names from Dalmatia, an Adriatic region that lies mostly within modern Croatia.

Of course, not all bacteria is species-specific. Dogs and humans can and do transmit some germs to each other via the mouth, so if your dog is the type that likes to lick faces (is there any other type?), there are a few precautions you can take. One, try to keep your dogs from picking up any external bacteria by keeping them out of the trash can (and away from rancid food) and away from wild animals (lest they contract rabies). Two, keep them healthy: up-to-date vaccines, good external and internal parasite control, regular teeth brushing, and the like. Once you do all that, pucker up!

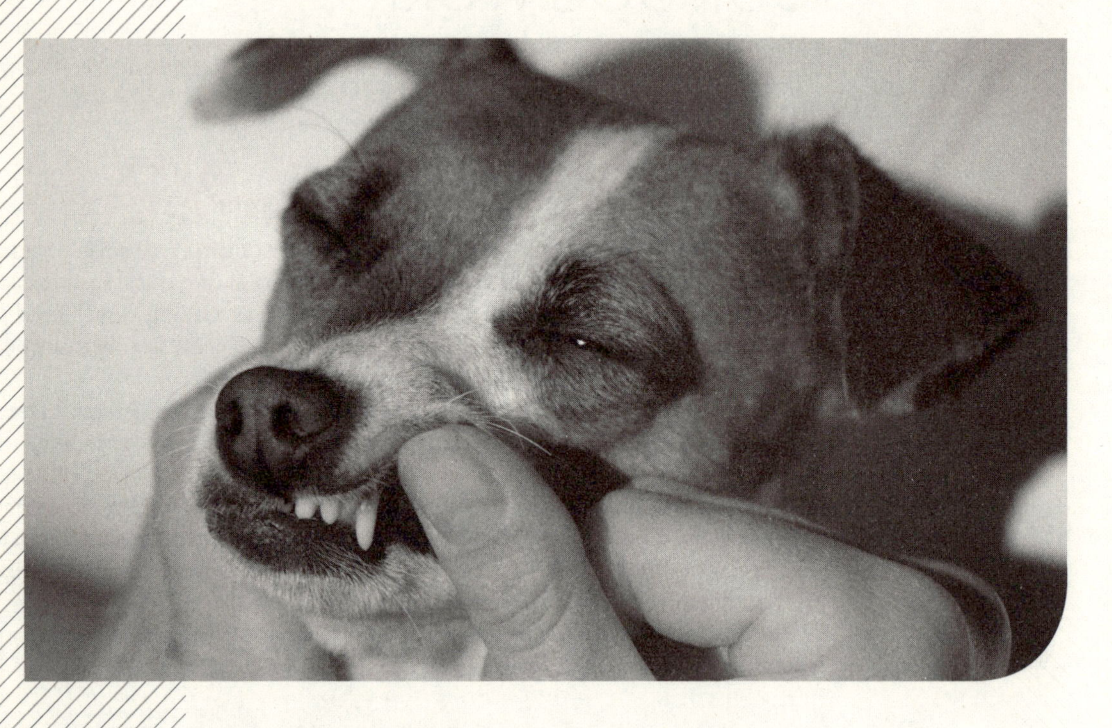

Why Don't Dogs Get As Many Cavities As People?

This one's simple. Cavities are largely the handiwork of the bacteria *Streptococcus mutans*. These bacteria feed on sugar, which is far more common in a human's diet than a dog's. Hence *S. mutans* prefers to live in our mouths, not Fido's.

Cocker spaniels got their name from English hunters. The little dogs were aces at tracking down woodcocks.

. . . AND ONE ANSWER FOR THE CAT LOVERS

What Is Catnip and Why Are Cats Crazy for It?

The secret to catnip is nepetalactone, a volatile oil stored in tiny bulbs on the leaves, stems, and seedpods of the plant. When nepetalactone enters a cat's nasal tissue, it binds to olfactory receptors at the olfactory epithelium. Sensory neurons are stimulated and cause neurons in the olfactory bulb to send signals to the brain. Scientists still don't have a complete neurological explanation for cats' behavioral reaction, but the prevailing theory is that nepetalactone mimics a cat pheromone.

Don't worry if your kitty's not cuckoo for catnip, though. It doesn't hold power over all felines. Response to catnip is genetically inherited, with about 70 to 80 percent of cats exhibiting the typical response to the plant. Of those, kittens younger than six months and very old cats are less likely to respond.

While catnip is related to marijuana, a 1988 *New York Times* article clearly stated, "[M]arijuana gives no pleasure to cats."

KEEP UNICORN HOPE ALIVE!

FIVE CREATURES PEOPLE DIDN'T THINK EXISTED

The Okapi

The Okapi was actually the mascot of the International Society of Cryptozoology because it was thought to be a myth until 1902. Zoologists even called it the African unicorn, which goes to show you just how outrageous people thought this animal was. Now you can find them pretty commonly in zoos.

The Kraken

The legendary beast sailors used to call the kraken may actually have been a giant squid. One account said the kraken was the size of a floating island, which would jibe with female giant squids' size. (They can be 43 feet long.) Scientists didn't get their first pictures of a fully grown giant squid alive until 2004, and in 2006, the first film was captured of a living, fully grown giant squid.

The Platypus

Just how unlikely did a platypus sound to scientists? When someone finally came forth with a body of everyone's favorite duck-billed, egg-laying mammal, the academic community assumed a prankster taxidermist had made the specimen by sewing parts of other critters together.

The Komodo Dragon

The Komodo dragons were mysteries until 1912 when a skin and a photograph of the "land crocodile" finally turned up. Scientists mounted an expedition to find more examples of the dragon and returned with 12 carcasses and two live specimens.

The Coelacanth

Talk about being left for dead. Scientists thought the coelacanth had been extinct for, oh, 65 million years or so before one showed up on the South African coast in 1938. Since then the giant fish—they average 176 pounds—have turned up in places as far flung as Tanzania and Indonesia.

The Quagga was a subspecies of zebra with stripes only across the front of its body. The last one died in captivity in 1883.

FOUR HORRIFYING PARASITES TO KEEP YOU AWAKE AT NIGHT

1 *Cymothoa exigua*: Biting Your Tongue, So You Don't Have To

When fish mommies want to strike fear in the hearts of their misbehaving fish babies, we suspect they draw on the chilling animal savagery of the *Cymothoa exigua*. As a youngster, this nasty little parasitic crustacean begins a life of terror by fighting its way through the gills of its fish host of choice, the snapper. Once there, it attaches itself to the fish's tongue and begins feeding on the rich blood pumping through the artery underneath.

As the parasite grows, it drinks more blood and eventually causes the tongue to atrophy and disintegrate. But does the *Cymothoa* mouth-squatter leave its fishy friend tongueless? Of course not. It does what any crafty parasite would do and replaces the old tongue with its own body. The fish is actually able to use the parasite just like a normal tongue, only it has to share all the food with its new friend. Yes, the whole foster-tongue thing seems like a pretty nice gesture on the part of ol' *Cymothoa*—until you remember there was nothing wrong with the fish's old tongue in the first place.

2 *Sacculina carcini*: Reasons You Shouldn't Pick Up a Hitchhiker

If you ever have a choice between being possessed by the devil and being possessed by a *Sacculina carcini*, opt for the devil—no contest. A female *sacculina* begins life as a tiny free-floating slug in the sea, drifting around until she encounters a crab. When that fateful day arrives, she finds a chink in the crab's armor (usually an elbow or leg joint) and thrusts a kind of hollow dagger into its body. After that, she (how to put this?) "injects" herself into the crab, sluicing through the dagger and leaving behind a husk. Once inside, the jellylike *sacculina* starts to take over. She grows "roots" that extend to every part of the crab's body—wrapping around its eyestalks and deep into its legs and arms.

The female feeds and grows until eventually she pops out the top of the crab, and from this knobby protrusion, she will steer the Good Ship Unlucky Crab for the rest of their commingled life. Packed full of parasite, the crab will forgo its own needs to serve those of its master. It won't molt, grow reproductive organs, or attempt to reproduce. It won't even regrow appendages, as healthy crabs can. Rather than waste the nutrients on itself, a host crab will hobble along and continue to look for food with which to feed its parasite master.

Only female mosquitoes will bite you.

3 *Leucochloridium paradoxum*: Parasite for Sore Eyes

Prepare to be dazzled. This parasite's got a life cycle more mind-bending and chilling than an M. Night Shyamalan film. *Leucochloridium paradoxum* are a type of fluke (a.k.a., parasitic flatworm) that prey on birds—a fascinating turn of events considering they begin their lives as eggs in bird droppings. Thus, the problem facing baby *Leucochloridium paradoxum* is, "How do I get myself back into one of those feathery things?"

Taking a page from Greek history, the infant flatworms rely on Trojan trickery. First, they hang out in the droppings until a snail happens along and eats the bird dung. Then they initiate their devious plan of action by taking up residence in the snail's eyestalks. (Sure, it sounds slimy and gross to us, but after a childhood spent living in bird feces, it's a step up.) As they mature, the flukes become visible through the snail's translucent skin. And that's when things really get interesting. To a bird, this fluke-filled eyestalk looks like a caterpillar. So the bird devours the stalk and ends up with a bellyful of *Leucochloridium paradoxum* that will, of course, lay eggs and begin the cycle again. Meanwhile, the snail shakes its head, shops for an eye patch, and vows never to eat feces again.

4 Filarial Worms: Proof You Need Thicker Skin

Filarial worms are the nasty little suckers you can thank for lymphatic filariasis, which, according to the Pacific Program to Eliminate Lymphatic Filariasis, is the second-leading cause of permanent and long-term disability in the world. (Mental illness is No. 1.) Filarial worms are round, thread-like parasites that travel from human to human via that harbinger of disease transmission, the mosquito.

How do they make the leap from host to host? In an interesting (if scary) example of parasite ingenuity, filarial worm embryos living underneath the skin can sense the onset of night, which is their cue to head upward to the skin's surface in order to increase their chances of being picked up by a passing 'skeeter. Should they get sucked up, they grow into larvae within the mosquito's muscle fibers and then get themselves injected into new hosts. Once they've returned to a human home, they open up a franchise in the family business—Wreaking Havoc. Filarial often lodge in the body's lymphatic system, where they can inflict any number of torturous symptoms, not the least of which involves carting your genitals off to the elephantiasis clinic in a wheelbarrow.

The 1900 Olympics featured a live pigeon-shooting event. Belgian Leon de Lunden won by bagging 21 pigeons.

TALKING TURKEY

EIGHT FACTS FOR THE TURKEY ENTHUSIAST

1 Butterball turkeys have been a holiday tradition for the past 50 years. While the name "Butterball" implies that the bird is injected with butter, it actually refers to a specific brand of turkey. Butterball turkeys have all-white feathers (birds with colored feathers often have unsightly dark spots on their meat) and have extra-broad breasts. Butterball turkeys are also the bestselling brand in the U.K. at Christmas, since the British don't celebrate Thanksgiving.

During Elizabeth I's reign, swan was a popular feast dish, especially when stuffed with the carcasses of nine other birds.

Caruncle, wattle, and snood might sound like a law firm, but they are actually words that describe the various bits of red fleshy stuff that grows on a turkey's head. The snood is the flap that hangs over its beak, the caruncles are colored growths on the throat, and the wattle is the skin hanging under the turkey's beak. When all three turn bright red the turkey is either in a mating mood or is very angry. In either case, you'll want to stay out of its way.

3

Unlike chicken and duck feathers, turkey feathers are too stiff for use as pillow and duvet stuffing. Some of the larger, more colorful feathers are sold for decorative purposes or craft projects, but the majority of turkey feathers are ground up and composted.

4

Some of those feathers that avoid the compost heap go on to big things, though. Big Bird of *Sesame Street* fame is clad in a costume made of turkey feathers plucked from the hind end of the bird. A company called American Plume and Fancy Feather selects the feathers for the Children's Television Workshop to inspect (nine out of 10 feathers are rejected). Then, the white feathers are dyed yellow and incorporated into Big Bird's costume.

5

The classic "Turkeys Away" episode of *WKRP in Cincinnati* was based on a very unsettling real event. *WKRP* creator Hugh Wilson had a friend who worked for an Atlanta radio station that decided to toss live turkeys out of a helicopter for a Thanksgiving promotion. Just like the TV episode revealed, none of the folks involved with the stunt knew that domestic turkeys couldn't fly, and a local shopping center was bombarded with turkeys hitting the ground "like bags of wet cement."

6

The folk melody "Turkey in the Straw" first gained popularity via minstrel shows in the mid-1800s. There is no copyright information available regarding the song, so the author of the tune remains a mystery. However, the song has earned at least one unusual place in pop history: In the United States, it is the most common melody used by ice cream trucks to attract customers.

7

It is now a Thanksgiving tradition for a live turkey to be presented by the National Turkey Federation to the U.S. president, and for him to grant it an official pardon. The lucky bird is then relegated to a farm or petting zoo to live out its life. For the record, John F. Kennedy was the first president to extend his pardoning power to poultry.

8

Even though domestic turkeys can't fly, their by-products are well-traveled. When *Apollo 8* astronauts tore into their Christmas dinner in space, their foil packets contained freeze-dried roast turkey with all the trimmings.

Benjamin Franklin on the turkey: "He is besides, tho' a little vain & silly, a Bird of courage, and would not hesitate to attack a (soldier) who should presume to invade his Farm Yard with a red coat on."

10 WORDS FOR ANIMAL PACKS

WEIRDER THAN A "MURDER" OF CROWS

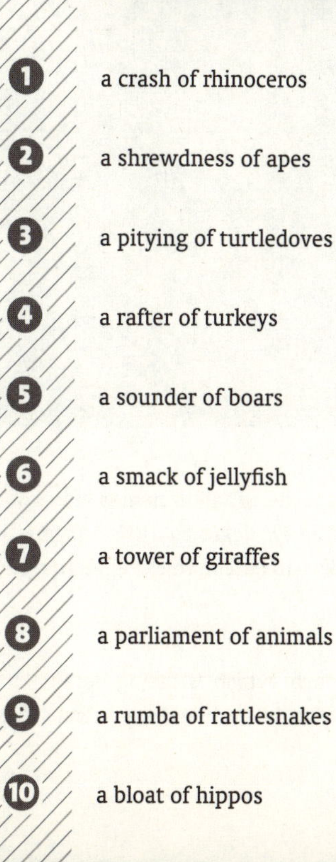

1. a crash of rhinoceros
2. a shrewdness of apes
3. a pitying of turtledoves
4. a rafter of turkeys
5. a sounder of boars
6. a smack of jellyfish
7. a tower of giraffes
8. a parliament of animals
9. a rumba of rattlesnakes
10. a bloat of hippos

Dalmatian puppies are actually born completely white. Their trademark spots only emerge two weeks after birth.

A STAR IS HATCHED

EIGHT ANIMALS NAMED AFTER CELEBRITIES

Sylvilagus palustris hefneri is an endangered species of rabbit named after someone else who adores the act of procreating: Hugh Hefner. The *Sylvilagus palustris hefneri* is a type of Marsh Rabbit that lives only in swampy areas and likes to hide in thick vegetation such as grass and bushes.

Cryptocercus garciai is a species of wood roach, which is fitting because it's named after Grateful Dead frontman Jerry Garcia. Nobody makes subtle drug references quite as well as zoologists!

About one in every 4 million lobsters is born with a rare genetic defect that turns it blue.

3. ***Phialella zappai*** is named after Frank Zappa simply because the guy who named that particular type of jellyfish wanted to meet Frank Zappa. Really. When Frank found out, he said, "There is nothing I would like better than having a jellyfish with my name," and invited the scientist to spend a couple of days at his house.

4. ***Calponia harrisonfordi*** and ***Pheidole harrisonfordi***. You don't get two creepy crawlies named after you without some work. Even if you are Indiana Jones. The scientist who discovered *C. harrisonfordi*, arachnologist Norman I. Platnick, named the spider to thank Ford for narrating a documentary for the London Museum of Natural History. The *Pheidole harrisonfordi* is a species of ant named after Ford to recognize his work as the Vice Chairman of Conservation International.

5. ***Preseucoila imallshookupis*** is a gall wasp. There's really no resemblance to Elvis; the scientist who named it was reportedly just an avid fan of the King.

6. The *Strigiphilus garylarsoni* wasn't the most glamorous animal to be named after *The Far Side*'s creator; it's a type of louse found only on owls. Larson mused on the name in *The Prehistory of the Far Side*: "I considered this an extreme honor. Besides, I knew no one was going to write and ask to name a new species of swan after me. You have to grab these opportunities when they come along."

7. ***Cirolana mercuryi***, a species of crustacean, was named after Freddie Mercury. Why? Because the little isopod makes its home in the coral reefs off of Bawe Island, Zanzibar. Since the Queen frontman was one of Zanzibar's most famous residents, the scientists thought it would be nice to honor Mercury.

8. ***Struszia mccartneyi*** puts Paul McCartney on the list. This one is a trilobite discovered in 1993. And lest you think that other aging rockers have been left out of the joke, don't worry—Mick Jagger, Johnny Rotten, Paul Simon, Art Garfunkel, Ringo Starr, and all of the Ramones also have trilobites named after them, as does John Lennon.

A goldfish-swallowing fad was so popular in 1939 that the *New York Times* published a story "Goldfish Gulpers Warned of Anemia."

THIS STUFF IS FOR THE BIRDS

SEVEN ANIMAL PHRASE ORIGINS

1 Scapegoat

If you're a scapegoat today, you're probably in deep trouble. But, ironically, the first scapegoats were lucky little buggers. In Biblical times, Hebrew priests would bring two goats to the temple on Yom Kippur. One goat, called the Lord's Goat, had a great-sounding title, but was saddled with the unfortunate duty of being the sacrificial victim. The other goat, called Azazel, was there to represent all the sins of the people. The priest would lead it back into the wilderness, symbolically taking the sins of all the Israelites with it. Heavy job, lucky goat. When European scholars began translating the Bible into English, they interpreted "Azazel" to be a variant of the Hebrew phrase, "the goat that departs." Or, the escape goat.

2 Pull the wool over their eyes

This phrase meaning "deceit" dates to the days when poofy, powdered wigs were all the rage. Whether or not you needed to join the Camaraderie of Tresses for Gentlemen, signing up was the only way to be accepted in polite society. Although this was probably the greatest social coup balding men have ever achieved, it did have its drawbacks. Mainly, it gave thieves a great way to distract a victim—by pulling his wig, or wool, over his eyes just long enough to run off with his wallet.

3 Red herring

So how exactly did this type of fish become synonymous with a false lead? The term goes back to when landed gentry in England would go on hunts in their private forests, and some creative interlopers figured out a way to get in on the goods. The trick: Getting between the hunting dogs and the prey they were chasing, and leading the pups off the scent by dragging a pungent red herring across the trail. Because trainers often used these fish to teach the hounds how to follow a scent, the dogs would follow the trail that smelled more familiar, leading the hunters away and giving the poachers a clear shot at the game.

4 Quack

While anyone who's ever used a See 'n' Say can tell you that, "The duck says quack," few can explain how duck dialogue became associated with unscrupulous (or just plain unskilled) medical practitioners. But the most likely solution stems from the fact that the q-word is what linguists call an imitative word. People heard ducks making noise, and "quack" was the closest they could get to the sound.

The nose print of a dog is like the fingerprint of a person—no two are alike.

Around the same time people decided that ducks quacked, they noticed another group that made a lot of noise: medicine salesmen. In the 16th and 17th centuries, the practice of medicine was largely in the hands of itinerant healers who traveled from town to town, hawking their salves, tonics, and ointments like carnival barkers. The cacophony of sales pitches reminded people of a flock of ducks, and the so-called healers became "quacksalvers," which was quickly shortened to "quacks."

5 Clotheshorse

Rendered somewhat obsolete in popular culture by words like "mallrat" and "valley girl," "clotheshorse" doesn't actually refer to the kind of horse you ride. Since at least 1391, the word "horse" has been used to describe any kind of four-legged device whose shape suggests an equine. Think sawhorse. Similarly, a clotheshorse was originally a wooden rack on which people hung their pants, shirts, and dresses out to dry. But by around 1850, the phrase was also being used to describe the sort of people who absolutely lived to show off their outfits.

6 Let the cat out of the bag

While a cat in a bag is probably never a particularly good thing, it was especially bad back in the days before grocery stores. In an attempt to dupe unwary shoppers, some devious farmers began switching out their customers' suckling pigs with cats. An earlier proverb associated with the same ruse, "never buy a pig in a poke," dates to 1562 ("poke" here being an old English term for bag). The cat itself wasn't let out, so to speak, until around 1760, when the phrase came to mean the revelation of any secret, not just your butcher's.

7 Hair of the dog

To take "a hair of the dog" refers to the questionable hangover remedy that prescribes a stiff drink the morning after a bender. But the phrase alludes to an even more questionable genre of remedies that were immensely popular in the Middle Ages. By the 16th century, any fool knew that the best way to treat an illness or injury was with a relic of whatever caused the malady. So, if a rabid dog bit a man, doctors would obtain some hairs from the dog and stitch them underneath the skin of the wound. As for the verbal connection between the cures for rabies and hangovers, it was first recorded in a collection of English proverbs from 1546, which quoted two gentlemen asking a bartender for "a hair of the dog that bit us last night."

Every thoroughbred horse is descended from one of just three "foundation sires."

10 LISTS TO READ BEFORE NAMING YOUR CHILD, COMPANY, OR ALTER-EGO

BEHAVE, BATMAN!

SIX BABY NAMES YOU PROBABLY SHOULDN'T GIVE YOUR KID

Forget Apple and Pilot Inspektor. If you really want to give your kid a hard time growing up, just pick from the following list.

1 Batman

Venezuelans are among the world's most creative namers. In fact, according to their own government, they're too creative. In September 2007, after hearing about babies named Superman and Batman, state authorities urged parents to pick their names from an approved list of 100 common Spanish monikers. Those conventional names (such as Juanita and Miguel) quickly acquired a patrician ring, ironically giving rise to more novel names, like Hochiminh (after the Vietnamese guerrilla) and Eisenhower (after the president). There are also at least 60 Venezuelans with the first name Hitler.

2 Eclipse Glasses

In June 2001, a total solar eclipse was about to cross southern Africa. To prepare, the Zimbabwean and Zambian media began a massive astronomy education campaign focused on warning people not to stare at the Sun. Apparently, the campaign worked. The locals took a real liking to the vocabulary, and today, the birth registries are filled with names like Eclipse Glasses Banda, Totality Zhou, and Annular Mchombo.

Michael J. Fox's middle name is Andrew.

3 Naaktgeboren

When Napoleon seized the Netherlands in 1810, he demanded that all Dutchmen take last names, just as the French had done decades prior. Problem was, the Dutch had lived full and happy lives with single names, so they took absurd surnames in a show of spirited defiance. These included Naaktgeboren (born naked), Spring int Veld (jump in the field), and Piest (pisses). Unfortunately for their descendants, Napoleon's last-name trend stuck, and all of these remain perfectly normal Dutch names today.

4 Vladimir Ashkenazy

The people of Iceland take their names very seriously. The country permits no one—not even immigrants—to take or keep foreign surnames. So what happened when esteemed Russian maestro Vladimir Ashkenazy asked to become an Icelandic citizen? Well, the government finally decided to make an exception. Vladimir Ashkenazy is now on the short list of approved Icelandic names.

5 Yazid

Imam Husayn ibn Ali is one of the holiest figures in the Shi'ite Muslim faith. In the 7th century CE, he lost his head on the orders of the Sunni caliph, Yazid, and the decapitation initiated the biggest schism in Islamic history. While the name Yazid remains common among Sunnis, it is disdained throughout the Shi'a world. The stigma attached to it is equivalent to naming one's son Stalin or Hitler. Speaking of which . . .

6 Adolf

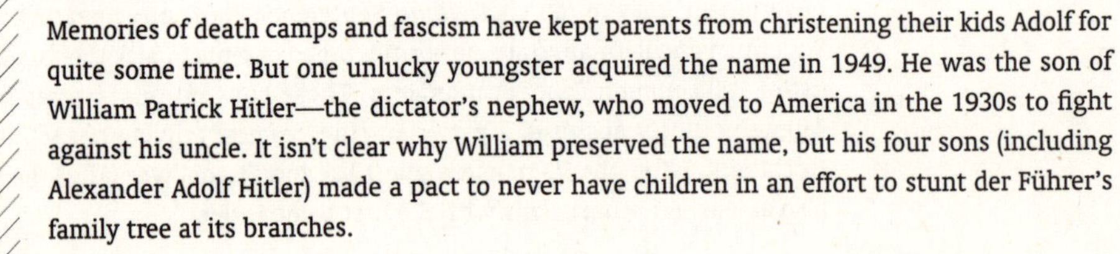

Memories of death camps and fascism have kept parents from christening their kids Adolf for quite some time. But one unlucky youngster acquired the name in 1949. He was the son of William Patrick Hitler—the dictator's nephew, who moved to America in the 1930s to fight against his uncle. It isn't clear why William preserved the name, but his four sons (including Alexander Adolf Hitler) made a pact to never have children in an effort to stunt der Führer's family tree at its branches.

Where do tough-sounding Leatherman tools get their name? A guy named Tim Leatherman started the company.

WHAT'S AN ATLANTA?

HOW SEVEN AMERICAN CITIES GOT THEIR NAMES

1 Portland

There was a 50-50 shot that Portland, Ore., was going to end up being called Boston, Ore. In 1845 what is now known as Portland was just a small settlement called "the Clearing." Settlers, Asa Lovejoy and Francis Pettygrove, both wanted to name the settlement after their own hometowns. Lovejoy was from Boston, while Pettygrove was from Portland, Me. The pair settled their argument by flipping a coin. Pettygrove and Portland won the best-two-out-of-three contest, and the city became Portland. The so-called Portland Penny they flipped is still on display at the Oregon Historical Society.

2 Atlanta

The ATL was very nearly the MAR. In the early 1840s, what is now Atlanta called itself "Marthasville," a nod to former governor Wilson Lumpkin's daughter Martha. The name changed to Atlanta in 1847, and although J. Edgar Thomson, chief engineer of the Georgia Railroad, gets credit for coining the "Atlanta" name, there is some debate over what inspired him. Some sources claim the aforementioned Martha Lumpkin's middle name was Atalanta. Others claim that Thomson took inspiration from Greek mythology's Atalanta. Still others claim that Thomson shortened the name from his original idea, "Atlantica-Pacifica."

3 Chicago

Chicago may be the Windy City, but its name has a fragrant origin. "Chicago" comes from the French pronunciation of *shikaakwa* the word for "wild garlic" in the Miami–Illinois language. Chicago was originally rife with the wild garlic we also know as ramps.

Cincinnati

Cincinnati was originally known as Losantiville, but that didn't sit well with territorial governor Arthur St. Clair. During a 1790 visit to Losantiville, St. Clair changed the name to Cincinnati to honor the Society of the Cincinnati, an organization of former Continental Army officers. (You guessed it; St. Clair was a member of the society.)

"Jay" used to be slang for "foolish person." So when a pedestrian ignores street signs, he's referred to as a "jaywalker."

Denver

Colorado's capital is named after James W. Denver, a 19th-century Renaissance man who served in Congress, fought in the United States Army, and served as Governor of the Kansas Territory. He only visited his namesake city twice, in 1875 and 1882, and was reportedly unhappy that the residents didn't give him more of a hero's welcome.

6 Phoenix

When the Arizona city was first taking off in the late 1860s, settlers realized that their little town needed a name. Founder Jack Swilling, a Confederate veteran, wanted to name the town Stonewall in honor of Stonewall Jackson, but Darrell Duppa recognized that their site had been a Native American settlement centuries earlier. He suggested Phoenix because their new city would rise from the ruins of the former civilization.

7 Cleveland

Cleveland takes its name from General Moses Cleaveland, a surveyor and investor for the Connecticut Land Company who led the first group to settle in the area in 1796. Cleaveland oversaw the planning of the early town before heading back to Connecticut a few months later. He never returned to the town that bears his name.

It's not exactly clear when the first "a" in his surname got dropped from the city's name, but one story explains that in 1830 *The Cleveland Advertiser* was short on space in its headline and simply axed the "a." The change caught on, and the town became known as Cleveland.

10 Common Words That Used to Be Trademarked

1. Cellophane
2. Crock-Pot
3. Dry Ice
4. Escalator
5. Heroin
6. Kerosene
7. Linoleum
8. Touch-Tone
9. Trampoline
10. Zipper

Alaska is the only state whose name can be typed on one row of keys. (Go ahead and try typing the other 49 states. We'll wait.)

A LOOK AT SHAGGY'S BIRTH CERTIFICATE

14 FICTIONAL CHARACTERS WHOSE NAMES YOU DIDN'T KNOW

1. Barbie's full name is Barbara Millicent Roberts. (Ken's last name is Carson.)
2. *Cap'n Crunch*'s full name is Captain Horatio Magellan Crunch.
3. In the Peanuts comic strip, Peppermint Patty's real name is Patricia Reichardt.
4. *Sesame Street*'s Snuffleupagus has a first name—Aloysius.

When 3-letter airport codes became standard, airports that had been using two letters simply added an X to their code, which explains LAX.

5. The real name of Monopoly mascot Rich Uncle Pennybags is Milburn Pennybags.

6. The Michelin Man's name is Bibendum.

7. On Gilligan's Island, Jonas Grumby was simply called the Skipper.

8. Mr. Grumby got to spend a lot of time with Roy Hinkley, better known as the Professor.

9. The unkempt Shaggy of *Scooby-Doo* fame has a rather proper real name—Norville Rogers.

10. The Pillsbury Doughboy's name is Poppin' Fresh. He has a wife, Poppie Fresh, and two kids, Popper and Bun Bun.

11. The patient in the classic game Operation is Cavity Sam.

12. Mr. Clean has a seldom-used first name—"Veritably." The name came from a "Give Mr. Clean a First Name" promotion in 1962.

13. Comic Book Guy on *The Simpsons* is named Jeff Albertson.

14. "The Wizard of Oz" rolls off the tongue a lot easier than his full name, Oscar Zoroaster Phadrig Isaac Norman Henkle Emmannuel Ambroise Diggs.

The tiny town of Ismay, Montana (population: 22), changed its name to Joe, Montana, for the duration of the 1993 NFL season.

YOU CAN MISLEAD A HORSE TO WATER

FOUR IRRITATINGLY INACCURATE NAMES

1 The Fertile Crescent

Though rich with several million years' worth of river silt, the land south and west of Baghdad, Iraq, is a not-so-fertile desert. From about 3500 BCE to the 15th century CE, the region's farms relied on a massive irrigation system. When the system failed, so did the agriculture.

White Chocolate

According to U.S. regulations, chocolate can contain no fat besides cocoa butter. White chocolate, however, is made with 30 percent vegetable fat, 30 percent milk solids, 30 percent sugar, and vanilla.

3 Allspice

It's not annoying that allspice doesn't contain all the spices, it's annoying that it doesn't contain any. In reality, allspice is just the dried berry of a single tropical American tree called Pimenta dioica.

4 The 100 Years' War

Historians rounded down. The war, a struggle between England and France (go figure) over the legitimate successor to the French crown, was fought off-and-on for 116 years, from 1337 to 1453.

Ten-gallon hat? Hardly. The average Stetson holds three quarts when filled to its brim.

MAKING A HUES DIFFERENCE

HOW SIX COLORS GOT THEIR NAMES

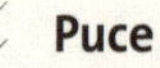

Puce

Puce means "flea" in French, and fleas are generally a reddish-brownish-purplish color.

Alice Blue

Teddy Roosevelt's eldest daughter, Alice, lent her name to this color that honors her grayish-blue eyes. The Navy uses Alice Blue as one of the colors on its insignia.

Fuchsia

Fuchsia takes its name from the flowers on the fuchsia plant. The plant is named for Leonard Fuchs, a 16th-century botanist.

Mountbatten Pink

Mountbatten Pink is the creation of Louis Mountbatten of the British Royal Navy. During WWII Mountbatten noted that a particular shade of pink was just about the color of the sky during dawn and dusk. He figured it would be ideal for camouflage during those hours.

Cerise

This one's easy—*cerise* is French for "cherry," which comes from the Norman *cherise*. "Hollywood cerise" is another name for the color "Fashion Fuchsia," which is a less-saturated version of fuchsia that's often used for clothing.

Cerulean

The color name has been around since at least 1677 and probably has Latin roots—*caeruleus* means dark blue, blue, or blue-green in Latin. In turn, *caeruleus* probably comes from the diminutive of the Latin word *caelum*, which means heaven or sky.

The poinsettia is named after congressman and ambassador Joel Poinsett, who introduced the plant to the United States in the 1800s.

THEIR NAMES ARE MUD

FIVE AWFUL EPONYMS (AND THE PEOPLE WHO CAN'T LIVE THEM DOWN)

1 Dunce

Dictionaries don't play fair, and John Duns Scotus is proof. The13th/14th-century thinker, whose writings synthesized Christian theology and Aristotle's philosophy, was considerably less dumb than a brick. Unfortunately for Scotus, subsequent theologians took a dim view of all those who championed his viewpoint. These "Scotists," "Dunsmen," or "Dunses" were considered hairsplitting meatheads and, eventually, just "dunces."

(Slipping a) Mickey

When you have to drug somebody against their will (hey, you gotta do what you gotta do), it just wouldn't sound right to slip 'em a Ricardo, a Bjorn, or an Evelyn. It's gotta be a Mickey. At the turn of the 20th century, Mickey Finn was a Chicago saloon owner in one of the seediest parts of town—and he fit right in. Finn was known for serving "Mickey Finn Specials," which probably included chloral hydrate, a heavy sedative. After targeted customers passed out, Finn would haul them into his "operating room" and liberate them of all valuables (including shoes). Never a Host of the Year candidate, this Mickey seems to have thoroughly earned his legacy, so don't hesitate to use it the next time you drug and rob your own customers.

The word "boycott" comes from Charles C. Boycott, an English landlord whose high rents led the entire region to stop selling him basic goods.

Spoonerism

Reverend William Archibald Spooner (1844–1930) was famous for his muddled one-liners. And though it's hard to know which ones he actually said, lines such as "I have a half-warmed fish" and "Yes indeed, the Lord is a shoving leopard" still prove that the sound-switching flub is pretty charming as far as mistakes go. The spoonerism has even been used as a literary technique by poets and fiction writers, giving Spooner little reason to roll over—or otherwise inarticulately protest—in his grave.

Shrapnel

While battling Napoleon's army, English General Henry Shrapnel (1761–1842) noticed that original-flavor cannonballs just weren't massacring enough enemies for his liking. So, to get more shebang for his shilling, he filled the cannonballs with bullets and exploding charges. These "shrapnel shells," or "shrapnel-barrages," were pretty darn effective, and later designs proved even more successful in World War I. Shrapnel didn't get much credit for the "innovation" during his lifetime, but he ultimately contributed to enough death and misery that he pretty much deserves to be synonymous with a violent, metallic by-product of combat.

Tawdry

The story of St. Audrey (also known as St. Etheldreda) is a classic example of how bad names happen to good people. St. Audrey was the daughter of the king of East Anglia (then the Norfolk section of Anglo-Saxon England), who lived a monastery-founding, self-abdicating life. But, when she died of the plague in 679, she was sporting a pretty nasty-looking tumor on her neck, which gossipmongers blamed on her penchant for wearing audacious necklaces in her youth. After her death, silk scarves called "St. Audrey laces" were sold in her honor at Ely's annual St. Audrey's Fair. Then the British tendency for dropping letters and syllables took over, and Audrey became "tawdry." It was a short trip from there to the dictionary, and tawdry has been synonymous with gaudy ever since.

The word chauvinism, originally meaning "exaggerated patriotism," comes from legendary French soldier Nicolas Chauvin.

A CORPORATION BY ANY OTHER NAME

SEVEN COMPANIES AND PRODUCTS THAT HAD TO CHANGE THEIR NAMES

1 Brad's Drink

By 1893 pharmacist Caleb Bradham had spent hours painstakingly formulating a new cola recipe to sell at his New Bern, North Carolina, shop. He then apparently spent two minutes naming it. Customers loved the sweet, fizzy tasty of "Brad's Drink," but the name didn't have much of a ring to it. Bradham went back to the drawing board and began calling his creation Pepsi Cola.

2 BackRub

In 1996, Stanford computer science grad students Sergey Brin and Larry Page started working on a new web crawling search engine. Since the engine used backlinks to gauge how important a site was, the enterprising pair called their creation "BackRub." By 1997 they decided this name wasn't so hot and changed it to "Google."

3 Jerry and David's Guide to the World Wide Web

"BackRub" sounds positively inspiring compared to this behemoth of a title. When David Filo and Jerry Yang started a guide to Internet content in 1994, they christened it "Jerry and David's Guide to the World Wide Web." Like Page and Brin, they quickly realized they might need a name that took less than three minutes to pronounce, so they switched to a word they liked from the dictionary—one that described someone who was "rude, unsophisticated, and uncouth." And that's how Yahoo! was born.

The I-Scream Bar

Danish immigrant Christian Kent Nelson came up with a brilliant snack while working at an Iowa confectionery store in 1919. After realizing that his young customers struggled with the choice between spending their allowance on chocolate or ice cream, Nelson perfected a process for coating a bar of ice cream with chocolate. He dubbed his creation the I-Scream Bar.

Denny's was originally called Danny's, but the name was too common to trademark. Substitute "e" to the rescue, and a diner empire was born!

By 1921 the I-Scream Bar had become so popular that Nelson needed to partner with a larger ice cream plant to mass-produce the treats. His new partners pointed out that an ice cream novelty aimed at children should probably have a less ominous name, so the I-Scream Bar became the Eskimo Pie.

5 Blue Ribbon Sports

In 1963, an ambitious young runner named Phil Knight met with the Japanese running shoe company Onitsuka about distributing their sneakers in the United States. The Japanese makers of Tiger running shoes decided to give Knight a shot, but they needed to know the name of Knight's company. It was a fair question to ask of a potential business partner, but Knight didn't actually *have* a company yet. Thinking on his feet, he quickly replied that he ran Blue Ribbon Sports.

Knight soon began selling Tigers out of his car at track meets around the United States. By 1971, though, his business had grown to the point where Knight was making his own shoes. He decided to name the shoes (and rename his business) after the Greek goddess of victory, Nike.

6 Cadabra

When hedge fund analyst Jeff Bezos began hatching a plan to launch an online bookstore in 1994, he was calling it Cadabra. Bezos soon realized that Cadabra sounded a bit too much like "cadaver," so he settled on a name that he felt summarized his store's sweeping volume and scope: Amazon.

7 ValuJet

ValueJet got its start in 1992 when former executives from the defunct Southern Airways joined up with former pilots, flight attendants, and mechanics from the defunct Eastern Airlines to start a new airline based in Atlanta. The airline showed early promise by offering low fares, but a disastrous May 1996 crash in the Florida Everglades left 110 people dead. In the wake of the tragedy, inspectors and pundits picked apart ValuJet's safety record, and confidence in the airline dwindled.

By the following summer, ValuJet looked to be on the ropes, but in July 1997 the company made an aggressive move. It bought AirTran Airways, an Orlando-based budget carrier, through a $61 million stock swap. Part of the deal involved changing the merged company's name to AirTran Airlines to get rid of the rotten public image of the ValuJet name. The move worked; when the airline officially dropped the tainted ValuJet name two months later, its stock price immediately shot up by more than 30 percent.

The 100 Grand Bar was the $100,000 Bar until 1985. Nestlé supposedly changed the name because it confused retailers' pricing computers.

CALL ME OPTIMUS!

FIVE CURIOUS PERSONAL NAME CHANGES

Not so fond of the name your parents gave you? You can always pull a Chad Ochocinco like these folks.

1 Schoolteacher Feels the Force

When *Star Wars Episode I: The Phantom Menace* hit theaters in 1999, one North Carolina woman managed to cash in on the hysteria. Sixth-grade teacher Jennifer Briggs heard a radio promotion that offered $1,000 to anyone who legally changed his or her name to Obi-Wan Kenobi in honor of the film's Jedi master. Briggs took the bait and became Obi-Wan Kenobi Briggs. Given her job, it might have been a brilliant decision; what sixth grader wouldn't listen to Obi-Wan?

On the 2001 New Zealand census, 53,715 people listed their religion as "Jedi."

Attempted Name Change Gets Dogged

In 2010 Pennsylvanian Gary Guy Mathews attempted to legally change his name to Boomer the Dog. Mathews enjoyed dressing as a dog and took the name from the short-lived 1980s NBC series *Here's Boomer.* According to the *Pittsburgh Post-Gazette*, Mathews was so serious about changing his name he even removed his dog collar for his day in court.

A Pittsburgh judge didn't think the name change was such a great idea, though. Judge Ronald W. Folino denied Mathews' petition on the logic that having a man named Boomer the Dog could lead to serious confusion. In his ruling, Folino noted that if Boomer the Dog had to call 911 in an emergency and identify himself, there was a decent chance the dispatcher would think it was a prank call.

3 Inventor Embraces His Inner Pronoun

Branson, Mo., inventor Andrew Wilson successfully changed his name to They in 2004. Just "They," no surname. The newly minted They admitted to the Associated Press that the name was a bit of a joke on people's tendency to refer to an abstract "they" in conversation. "'They do this,' or 'They're to blame for that.' Who is this 'they' everyone talks about? 'They' accomplish such great things. Somebody had to take responsibility," They said.

4 New Name Transforms Man

When Ohio's Scott Nall turned 30, he decided he was tired of being Scott Nall. As a birthday present to himself, he legally changed his name to Optimus Prime, the lead protagonist of the Transformers line of toys and movies. Optimus the guy had more than a little in common with Optimus the brave robot, too; he has served as a firefighter in Iraq.

Ukrainian Politician Chooses Default Name

In 2009 Ukrainian pensioner Vasyl Humeniuk decided he'd had enough of his country's politicians, so he changed his name to Vasyl Protyvsikh. (For the few readers who don't speak Ukrainian, the new surname translates into "Against-Everyone" or "None of the Above.") Protyvsikh then announced that he was entering the country's upcoming presidential election.

Offering voters a chance to cast a ballot for "None of the Above" is an inspired political ploy, but it didn't work out so well for Protyvsikh. He failed to pick up even one percent of his country's votes.

"Velcro" is a combination of the words "velvet" and "crochet," the French word for "hook."

OUR COMPLIMENTS TO CHEF BOYARDEE

FIVE PRODUCTS NAMED AFTER REAL PEOPLE

1 Duncan Hines

Today's travelers rely on Frommer's or the Michelin guide when it comes to restaurant ratings, but in the 1940s, a recommendation by Duncan Hines was the ne plus ultra statement in fine dining. Duncan Hines was a traveling sales representative for a Chicago printing company during the 1930s and 1940s, long before the advent of chain restaurants. To help quell the boredom of those long, pre-Interstate road trips, Hines kept a diary of the restaurants where he'd dined along the way.

In 1935, in lieu of a Christmas newsletter, he and his wife sent a list of 167 restaurant reviews they'd compiled to friends and family. Hines was eventually approached by a book publisher looking to capitalize on his gustatory expertise, and later by a manufacturer of pre-packaged foods looking to use his prestigious name on their products.

Chef Boyardee

Ettore "Hector" Boiardi was born in Piacenza, Italy, and began working as a cook in a local restaurant at the tender age of eleven. He was 16 when his family moved to the United States, and one year later he landed the job of Head Chef at New York's prestigious Plaza Hotel. When President Woodrow Wilson married Edith Galt in 1915, Boiardi catered the event.

In the early 1920s Boiardi accepted the position of Head Chef at Cleveland's Hotel Winton, where he featured a menu that emphasized the traditional Italian cuisine he so loved. It wasn't long before patrons started asking Chef Boiardi for his spaghetti sauce recipe, which he refused to share. He did, however, sell it as a "to go" item, using empty milk bottles as containers. He opened his own restaurant in 1924, and when the volume of take-out orders surpassed the number of sit-down customers, he opened a separate factory that packaged his products for sale in retail outlets. He used a phonetic spelling of his surname on the product labels so that there was no confusion as to how it was pronounced.

3 CliffsNotes

Some of us would never have made it through the literary classics without the help of Clifton Hillegass. He is the "Cliff" behind CliffsNotes, those little yellow study guides that condense a hundred pages of Shakespeare into three concise paragraphs. Hillegass was a graduate of the University of Nebraska and an Army Air Corps veteran. After World War II, he got a job as the manager of the wholesale department of the Nebraska Book Company, a textbook store.

Despite the Japanese name, Ginsu Knives were originally manufactured in Springfield, Ohio.

When Hillegass published his first Cliff's Notes (the series eventually lost the apostrophe) in the basement of his Lincoln home, it was with the intent of enriching the reader's experience and pointing out plot subtleties, not providing a "cheat sheet." In fact, each volume of his study guides included a signed note to his readers that stated: "A thorough appreciation of literature allows no shortcuts." Hillegass started his company with a $4,000 loan in 1958 and published 16 Shakespeare study guides. In 1998, he sold CliffsNotes to IDG Books for $14 million. "Cliff" passed away in 2001, but his memory lives on in his native Nebraska. For 40 years, he donated 10 percent of his profits to local charitable causes.

4 Vicks Vapo-Rub

After graduating from college, Lunsford Richardson became a professional druggist and worked in a pharmacy owned by his brother-in-law, Dr. Joshua Vick. Richardson spent much of his spare time in the back room of the drugstore concocting various home remedies, such as headache powders and liniment.

One winter all three of his children caught bad colds, and his wife used the traditional treatment of that time—placing a poultice on the patient's chest and then lighting a kerosene vaporizer lamp. Richardson thought that there might be a way to combine the two—poultice and vaporizer—that would provide more immediate relief to cold symptoms. He eventually created a mixture of menthol, camphor, oil of eucalyptus, and petroleum jelly that sold itself once the first few customers tried it. When his invention started selling on a national scale, he decided that a short, easy-to-remember name was needed, so in honor of his brother-in-law, he christened it Vicks Vapo-Rub.

5 Maybelline

'Twas unrequited love that inspired a cosmetics empire. Chicago chemist Thomas Williams had an older sister named Maybel, and in 1913 Maybel had a major crush on a man who was in love with someone else. She did what she could to make herself more appealing than her competitor, including applying petroleum jelly to her eyelashes and eyebrows to enhance them. Tom wanted to help his sister get her man (or perhaps he was just tired of her crying on his shoulder) and went to work in his laboratory. He came up with a formula of carbon dust added to petroleum jelly which, when applied to the lashes and brows, highlighted them dramatically.

Two years later, Maybel married the object of her affection, and Thomas found out there was serious money to be made when it came to helping women look their best in an era when marrying well was the loftiest dream to which a female could aspire. He started selling eye makeup under a name inspired by his sister, "Maybelline"; his products were originally sold by mail order only but due to overwhelming demand he eventually moved his wares into retail stores.

The duffel bag gets its name from the town of Duffel, Belgium, where the cloth used in the bags was originally sold.

10 LISTS OF LEMONS

THE FIVE MOST ANNOYING PEOPLE IN HISTORY

1 Guglielmo Marconi: How Bribery Killed the Radio Star

The Good: Winning the Nobel Prize for inventing radio-transmission technology.
The Totally Infuriating: Winning the Nobel Prize for inventing radio-transmission technology when you aren't the guy who invented it. In the early 1890s, Nikola Tesla discovered (and widely published) the means to transmit and receive radio signals using high-frequency electric currents. The next year, Guglielmo Marconi took out an English patent on the same system.

A remarkably good sport, Tesla decided to simply patent his work in America. Then Marconi tried to do the same thing. At first, the U.S. Patent Office resisted, noting that Marconi's claim to know nothing of Tesla's work was more than a little ridiculous. But Marconi had something Tesla didn't—connections. His family ties to English aristocracy helped the Marconi Wireless Telegraph Co. thrive and led to his friendships with American luminaries such as Thomas Edison.

The original duty of a wedding's "Best Man" was to serve as armed backup for the groom in case he had to resort to kidnapping his intended bride.

In 1904, Marconi's networking paid off. The U.S. Patent Office suddenly changed its mind. For reasons never fully explained, Tesla's patent was voided and given to Marconi, who became rich and won the Nobel. Poor ol' Tesla died, well, poor. But the story doesn't end there. In 1943, the Patent Office switched sides again and gave the radio patent back to Tesla.

2 Caravaggio: That Guy Who Makes John McEnroe Look Calm

Caravaggio was a 16th- and 17th-century Italian painter who's best known for creating the technique of tenebrism, or dramatic illumination. But he was also a world-class thrower of temper tantrums—long after he was old enough to know better. Famous for starting fights all over Rome, Caravaggio had a rap sheet a mile long. His offenses included the mundane (attacking another artist, 1600), the comical (throwing a plate of artichokes in the face of a waiter, 1604), and the just plain dumb (killing a man over a disputed tennis score, 1606).

3 Biagio da Cesena: Patron Saint of Bad Art Critics

As Papal Master of Ceremonies for Pope Paul III in the mid-16th century, Biagio da Cesena had a legitimate reason to be interested in the frescoes Michelangelo was painting in the Sistine Chapel. But he didn't have to be so nosy about it. Da Cesena spent four years bugging Michelangelo to give him a sneak peek at the work, and then, when he finally did see it, he tried to have the art destroyed. Michelangelo had included nude figures in some of the scenes, and da Cesena called them a disgrace.

Fed up with the Master of Ceremonies' whining, the artist painted da Cesena into one of the circles of hell. In response, da Cesena did the mature thing and told the pope that Mikey was picking on him. But the pope was apparently not fond of tattling. "God has given me authority in heaven and on Earth," he declared, "but my writ does not extend to hell."

When Henry Budd died in 1862, he left his substantial fortune to his two sons on the condition that neither sullied his lip with a mustache.

Pete Best: That Guy Who Just Won't Let It Go

In 1962, drummer Pete Best was fired by the Beatles in favor of Ringo Starr. Since then, he's embarked on a series of embarrassing and absurd displays of ill-affection that make him the biggest sore loser in the history of the universe. Cases in point: (1) getting into a fight with the Beatles during which a rabid Best fan gave George Harrison a black eye; (2) suing Ringo Starr for libel (Starr told *Playboy* that Best was fired because of drug abuse); (3) suing Trivial Pursuit for libel (they accidentally claimed Best was dead); and (4) capitalizing off the Fab Four well into old age with his deceptively named album *Best of the Beatles*.

Arthur Schopenhauer: One of Those "Glass Half Empty" Kind of Guys

It's probably safe to say that a man known as "the philosopher of pessimism" would be kind of a downer at parties. Enter Schopenhauer, whose writings described life as a continuous toil toward death. Obsessive about his routines, Schopenhauer began each day by reading the paper to become "informed of the world's miseries." He spent his afternoons teaching and would purposely schedule his lectures at the same time as those of competing philosopher Hegel—just to see which instructor students would choose. In the evening, Schopenhauer either went to the opera (where he would angrily confront latecomers and "loud-coughers") or whiled away the hours insulting women down at the local pub.

The infamous Venus flytrap lives only in one tiny swath of natural habitat—a 60-mile radius around Wilmington, North Carolina.

GIVE US YOUR TIRED, YOUR POOR, YOUR DEAD SEAHORSES

FIVE ODD THINGS CUSTOMS HAS SEIZED

Most of us feel like outlaws if we sneak an extra bottle of duty-free booze past Customs, but some smugglers eschew importing drugs, guns, and Cuban cigars to set their sights higher.

1. In 2010, French Customs cracked the case of 354 counterfeit Fabergé eggs headed to the country from Russia. If nothing else, the sheer number of the jeweled beauties should have tipped them off: only 42 original Fabergé eggs are known to still exist.

2. Eighty pounds of dead seahorses failed to make it through New Jersey's Elizabeth Seaport in 2009. You might wonder why anyone would bother to smuggle stiff seahorses: well, rumor has it the little guys bring $800 an ounce in some markets because they can be used as the Viagra of the Sea.

3. What's wrong with importing 316,000 Christmas ornaments? Nothing, unless the ornaments in question are actually bongs. That was the case in a 2009 bust when $2.6 million in dubious "Christmas ornaments" and "glass sculptures" turned up at the Los Angeles harbor.

4. Looking to cash in by discovering a priceless lost painting by one of the greats? Forget scouring attics and garage sales—just check Customs. Polish Customs officers nabbed a previously unknown painting by Renoir before it could be shipped to the United States in early 2011.

5. U.S. Customs seized 250 copies of Allen Ginsberg's poem *Howl* on March 25, 1955, refusing to allow the English shipment in on the grounds that the book was obscene. The setback didn't stop Ginsberg—a San Francisco publisher picked it up the following year and it became widely distributed in the United States.

General Douglas MacArthur's mom rented a hotel room across the street from her son's dorm at West Point to keep tabs on his study habits.

CENTI-WHAT?

SIX UNIT CONVERSION DISASTERS

1. Can you imagine losing $125 million thanks to a little metric system error? That's exactly what happened in 1999 when NASA lost a Mars orbiter because one team used metric units for a calculation and another team didn't. Guess they didn't learn from their previous mistake . . .

2. . . . just the year before, NASA lost equipment worth millions thanks to shoddy conversion practices. SOHO, the Solar and Heliospheric Observatory, a joint project between NASA and the ESA (European Space Agency), lost all communications with Earth. After about a week of trying various things, communications came back online and everyone breathed a sigh of relief. Among the problems thought to have caused the sudden blackout? Someone had forgotten to put English-to-metric conversion factors into the observatory's files.

3. In 1983, an Air Canada plane ran out of fuel in the middle of a flight. The cause? Not one but two mistakes in figuring how much fuel was needed. It was Air Canada's first plane to use metric measurements, and not everyone had the hang of the conversions yet. Luckily, no one was killed and only two people received minor injuries, which is amazing considering the flight crew thought they had double the fuel they actually had.

4. In 1999, the Institute for Safe Medication Practices reported an instance where a patient had received 0.5 grams of phenobarbital (a sedative) instead of 0.5 grains when the recommendation was misread. (A grain is a unit of measure equal to about 0.065 grams. Yikes!) The Institute emphasized that only the metric system should be used for prescribing drugs.

5. An aircraft taking off more than 30,000 pounds overweight is no laughing matter. In 1994, the FAA received an anonymous tip that an American International Airways (now Kalitta Air, a cargo airline) flight had landed 15 tons heavier than it should have. The FAA investigated and discovered that the problem was in a kilogram-to-pounds conversion (or lack thereof).

6. The other inept converters on this list should know that they're in good company. Even Columbus had conversion problems. He miscalculated the circumference of the earth when he used Roman miles instead of nautical miles, which is part of the reason he unexpectedly ended up in the Bahamas on October 12, 1492, and assumed he had hit Asia. Whoops.

Only 3 countries haven't adopted the metric system as the official system of weights/measures—Burma, Liberia, and the United States.

BONEHEADS

FOUR MASSIVE SCREWUPS IN PALEONTOLOGY

Thomas Jefferson's Giant Mistake

In decades past, American presidents apparently had hobbies other than playing golf. Thomas Jefferson, for one, was an avid paleontologist. As early as the 1790s (before it was cool), he kept an impressive fossil collection at his home in Monticello. So when a group of confused miners came upon some unidentifiable bones in a West Virginia cave, they sent them to Jefferson. Judging from the long limbs and large claws, the president suspected they belonged to a giant cat "as preeminent over the lion in size as the mammoth is over the elephant," and that the animal might still exist somewhere in the unexplored West.

Jefferson got the size right. The description? Not so much. The animal he named Megalonyx (giant claw) was actually one of the giant ground sloths that slowly roamed America during the last ice age. And while Jefferson later agreed with this alternative diagnosis, his error wasn't a complete waste. The Megalonyx marked one of the first important fossil finds in the United States, and it prompted the first and second scientific papers on fossils published in North America. In honor of the president's contribution, the sloth's name was later formalized to *Megalonyx jeffersonii*.

Stephen Hawking revealed in a 2006 lecture that Pope John Paul II had discouraged the scientist from studying the beginning of the universe.

2 The Dinosaur That Never Was

To this day, the Brontosaurus remains one of the most popular and recognizable dinosaurs in history—an impressive feat for an animal that never existed. The confusion started in 1879, when collectors working in Wyoming for paleontologist Othniel Charles Marsh found two nearly complete—yet headless—sauropod dinosaur skeletons. Wanting to display them, Marsh fitted one specimen with a skull he found nearby, and the other with a skull he found in Colorado. Voilà!—the Brontosaurus was born.

Unfortunately for Marsh, the skeletons were later exposed as adult specimens of a dinosaur he'd already discovered, the Apatosaurus. The error was formally corrected in 1903 by Elmer Riggs of Chicago's Field Museum, and scientific papers haven't called the animal Brontosaurus since. Seventy more years passed before researchers determined that the skulls Marsh borrowed really belonged to the Camarasaurus, a discovery of his archrival, Edward Drinker Cope. Pop culture, however, missed the memo altogether.

Pulling Teeth

Henry Fairfield Osborn was a giant in the field of paleontology, but he also has one giant mistake to his name. In 1922, while serving as president of the American Museum of Natural History, Osborn received a fossil of a tooth found in Nebraska. Suffering from a bout of overconfidence, the normally careful scientist published a paper announcing (based on one tooth, mind you) that he'd discovered *Hesperopithecus haroldcookii*, the first anthropoid ape unearthed in North America.

Taking into account that all of this was happening just three years before the Scopes Monkey Trial, word of a missing link was a pretty big deal. Add to that British anatomy professor Sir Grafton Elliott Smith touting the discovery as a potential breakthrough, and artist Amedee Forestier drawing a famously speculative picture of the "Nebraska Man" (and Woman) in the widely read *Illustrated London News*.

Although Osborn never hypothesized where (or if) his ape fit into the evolutionary chain, he used the discovery to fuel his war of words with antievolution blowhard William Jennings Bryan. Osborn made sure to note the irony of the tooth having come from Bryan's home state, and even suggested calling the ape *Bryopithecus* in honor of "the most distinguished primate which the state of Nebraska has thus far produced."

Unfortunately, in this particular case, said distinguished primate got the last laugh. Upon further examination, it was determined that the tooth belonged to a millennia-old peccary—otherwise known as an ancient pig. In fairness to Osborn, the similarities between human and peccary teeth had already been noted in scientific literature, so it wasn't that wild a guess. Of course, that didn't stop creationists from pouncing on the mistake.

During a 2009 search for the Loch Ness Monster, scientists in Scotland found over 100,000 submerged golf balls instead.

Creating a Monster

Long before there was a science called paleontology, people were trying to come up with explanations for giant bones found in the ground. And often, those explanations pointed to mythological creatures. But of all the fairy-tale creatures accused of inhabiting the ancient world, the griffin might claim the most direct connection to actual fossils. Usually depicted in folklore as a lion with an eagle's head and wings, the griffin was said to fiercely guard its gold. The hybrid animal appears consistently in the art of ancient Rome, Greece, and Persia, and its legend apparently originated with Scythian nomads who wandered east toward Mongolia's Gobi desert.

So, how do fossils fit in? The Gobi is filled with the fossils of both the Protoceratops, a lion-size dinosaur with a birdlike beak, and of the similarly beaked Psittacosaurus. And while there were no massive hoards of gold around, the skeletons were often found guarding something arguably more valuable—hoards of eggs. The ancients were wrong about griffins, but that may have had more to do with misdiagnosing evidence than with legend or superstition.

The "black box" on an airplane is actually blaze orange, which makes it easier to find amid wreckage after a crash.

THE ART OF WAR

THREE ARTISTIC RIVALRIES THAT TURNED UGLY

Voltaire Exposes Rousseau As a Deadbeat Dad (Five Times Over?!)

They say you're not paranoid if someone's really out to get you. Jean-Jacques Rousseau was paranoid, but Voltaire was really out to get him, too. The two philosopher/writers started sniping at each other in the 1750s. At the time, Voltaire was an established leader of French philosophical circles and Rousseau (yet to write *The Social Contract*) was just a newbie. But the balance of power began to shift when Voltaire moved to Rousseau's native city of Geneva in 1754.

Although Rousseau had left Geneva in 1728, he remained devoted to the city's strict Calvinist standards, which included a ban on public plays. So when he heard Voltaire was not only putting on private dramas but also urging city authorities to admit plays into the city, Rousseau wrote an outraged letter condemning theatricals. In return, an annoyed Voltaire wrote to his philosopher friends saying that Rousseau had only criticized the theater because Rousseau had written a bad play.

Rousseau went off the deep end. He dipped his pen in vitriol and scratched out a letter to Voltaire that began, bluntly enough, "I do not like you, sir." He went on to outline all the (perceived) slights he'd received from Voltaire and concluded, "In a word, I hate you." Voltaire thought Rousseau had lost his mind and publicly advised his fellow philosopher a course of soothing baths and restorative broths.

Henceforth, Voltaire would miss no opportunity to slam his enemy. He mocked the plots of Rousseau's novels, insinuated Rousseau had inflated his résumé, and bashed Rousseau's book *Julie* as "silly, middle-class, dirty-minded, and boring." Finally, in 1764, Voltaire wielded the most powerful weapon he possessed—a secret about Rousseau he'd picked up in Geneva. Using a pseudonym, Voltaire wrote an open letter accusing Rousseau of abandoning his five children at the door of an orphanage. The accusation was shocking—and true.

Rousseau, in a politician-worthy statement of denial, could only claim, "I have never exposed, or caused to be exposed, any infant at the door of an orphanage." He was telling the truth, but only because the children had been taken inside the orphanage. Further scrambling to justify his actions, Rousseau responded with his book *Confessions*, now recognized as one of the first true autobiographies. An ugly quarrel, it seems, marked the invention of a new literary form.

In 1999, the U.S. government paid the Zapruder family $16 million for the film of JFK's assassination.

2 Dueling Pianos: Johann Mattheson Almost Kills George Frideric Handel

Johann Mattheson met fellow composer George Frideric Handel in 1703, when the 21-year-old Handel moved to Hamburg to take the position of violinist and harpsichordist for the opera-house orchestra. This made Handel something of a young celebrity, but Mattheson was something of a celebrity himself, being a former child prodigy and a popular local composer. The two hung out quite a bit, and Mattheson even gave Handel advice on writing his first opera.

But the friendship hit a roadblock in December 1704, when Mattheson premiered his third opera, *Cleopatra*. Mattheson not only wrote and conducted the piece but also sang the part of Antonius. (Busy guy, Johann.) During the first three-quarters of the performance, Mattheson was onstage. But half an hour before the end, Antonius commits suicide, which left Mattheson at loose ends. Deciding to take over at the harpsichord, he headed for the orchestra pit and whispered to Handel, who was then tickling the ivories, to scoot over. A very much miffed Handel refused to give way.

History doesn't note the effect of the musicians' brawl on the performance, but it does record that Mattheson challenged Handel to a duel. According to Mattheson, the two retired to the street, took up swords, and started slashing. Also according to Mattheson, his sword broke when it struck one of Handel's large metal coat buttons, which is the only reason George's life was spared.

In 1969 George Harrison quit the Beatles for five days. John Lennon's reaction: "If he doesn't come back by Tuesday, we get Eric Clapton."

Either way, Handel went on to bigger and better things (*Messiah*, for one), while Mattheson remained in Hamburg churning out oratorios. Watching Handel's rise from afar, he once complained that Handel had stolen the melody from one of his operas. (Probably a true charge, actually, as Handel was notorious for "borrowing" melodies.) Finally, toward the end of his life, Mattheson filled his autobiography with stories of his world-famous buddy, taking as much credit as possible for himself.

3 Lillian Hellman and Mary McCarthy Duke It Out

One January night in 1980, playwright Lillian Hellman (*The Little Foxes*) sat up in bed while watching *The Dick Cavett Show*. Novelist and critic Mary McCarthy was on the program discussing books when Cavett asked her which writers she considered overrated. "Lillian Hellman," McCarthy promptly replied. "Everything . . . every word she writes is a lie, including 'and' and 'the.'"

Hellman may have been 74 years old, nearly blind, and unable to walk, but she could still use the telephone. She called her attorney and ordered him to sue McCarthy—along with Cavett, the show's producer, and the station—for $2.25 million in libel. The result was a public slugfest, with all of America's writers taking sides. Norman Mailer tried to act as a peacemaker via an article in the *New York Times*, but it only proved to annoy both sides. Hellman even offered to drop the suit if McCarthy publicly apologized, to which McCarthy responded, "But that would be lying."

To the surprise of everyone, including Hellman's attorneys, the New York Supreme Court refused McCarthy's request to dismiss the case on May 10, 1984. Sadly, Hellman didn't have long to enjoy her victory; she died less than two months later. McCarthy, who'd been facing financial ruin, was less than satisfied, complaining, "I didn't want her to die. I wanted her to lose in court." Since then, the case has been remembered in legal circles as raising important free speech issues. As *Harper's* magazine quipped, "If you can't call Lillian Hellman a liar on national TV, what's the First Amendment all about?"

Jakob Dylan, lead singer of The Wallflowers, stipulates in his contract that no advertiser or interviewer should ever refer to him as "Bob Dylan's son."

KEEP TO YOURSELF!

FOUR INFAMOUS QUARANTINES

1 Irish Cook Gets Her Own Island, Infamous Nickname

Irish immigrant Mary Mallon was originally just another cook and caretaker in the New York City area in the early 1900s. Then, families developed a habit of suddenly becoming stricken with typhoid even though Mallon herself appeared to be perfectly healthy.

In 1915 doctors ordered that Mallon be quarantined for life after determining that she had infected more than 30 people, three of them fatally. Although her status as a healthy carrier of the disease earned her the nickname "Typhoid Mary," she didn't die of typhoid. Instead, she passed away in 1938 following a stroke after spending over two decades quarantined on North Brother Island in New York's East River.

2 It Takes a Village to Raise a Quarantine

When the English village of Eyam received a bolt of flea-ridden cloth in 1665, residents knew they had a potential plague problem on their hands. Villagers soon began falling victim to the plague, so the town took the drastic step of quarantining itself to slow the disease's spread. The quarantine lasted for over a year, during which time over half of the village's population succumbed to the Black Death.

You Can't Quarantine Racism

In 1900, the Chinese owner of a lumberyard in San Francisco died of the bubonic plague. In a classic case of overreaction, authorities immediately roped off 15 blocks surrounding the area, which included 25,000 people of Chinese descent and their businesses. Shops owned by Caucasian people were not subject to the seclusion, and after three months a court ruled the quarantine was unfair and racist. The court declared health officials used an "evil eye and an unequal hand."

You Guys Are Heroes! See You in a Few Weeks . . .

The crew of the *Apollo 11* mission that landed on the moon received a booming welcome when they returned home in 1969. Buzz Aldrin, Neil Armstrong, and Michael Collins then went into quarantine for three weeks to ensure that they hadn't brought any space bugs back from the lunar surface with them. The astronauts of *Apollo 12* and *Apollo 14* got the same treatment before scientists determined that the moon was devoid of all life, including germs.

In 2000 the *New York Times* reported that Queen Elizabeth II had a Big Mouth Billy Bass on her grand piano at Balmoral Castle.

DRESSED TO KILL

FIVE ARTICLES OF CLOTHING THAT CAUSED RIOTS

The clothes may make the man, but sometimes it's what the clothes make the man do that makes the story.

A Top Hat

In 1797, London haberdasher John Hetherington was hauled into court on charges of breaching the King's peace, found guilty, and ordered to pay a £50 fine. His crime? Wearing a silk top hat, or, as it was described in court, "appearing on the public highway wearing upon his head a tall structure having a shining lustre and calculated to frighten timid people." According to contemporary reports, people booed, dogs barked, women fainted, and a small boy suffered a broken arm after a crowd formed around the hapless Mr. Hetherington.

After protests in 2005, Vermont Teddy Bear discontinued its "Crazy for You" bear, which came with a straitjacket and commitment papers.

Top hats were evidently outlawed in London for a time after that, although not for very long—50 years later, Prince Albert boosted the hat's popularity in England by wearing one, and establishing the primacy of the top hat for generations to come. In America, it's virtually impossible to picture Abraham Lincoln without it, Monopoly just wouldn't be the same without it, and what else would Uncle Sam possibly wear?

2 Straw Hats

Acceptance of the top hat grew, and by the 1920s, women didn't faint and dogs didn't bark at seeing gentlemen attired thusly. But the straw hat, however, is a different story.

Over several nights in September 1922, gangs of hundreds of young thugs terrorized Manhattan, destroying any "unseasonable straw hat" they found. According to contemporary *New York Times* reports, these fashion vigilantes were armed with sticks, some with nails at the ends, and forced men in straw hats to run "gauntlets" of fists and boots. The streets were littered with broken and trampled straw hats and the remains of straw hat bonfires. The police were called in to disperse the unruly hat-haters, and hat stores were forced to stay open late to accommodate the newly hatless.

According to the *Times*, the hat-smashers were gangs of mostly young boys who took the September 15 end of straw hat season very seriously. While Magistrate Peter A. Hatting (no, really) upheld the inalienable right of a man to wear a straw hat "in a January snowstorm if he wishes," the hat-smashers disagreed, choosing instead to attack any straw-hatted person and destroy his hat for him. Dozens were arrested and fined over the course of the riots and people, including several off-duty and presumably straw hat–wearing police officers, were injured.

Oddly, the same scenario had unfolded only 10 years earlier, in Bridgeton, New Jersey, when the official end of hat season was September 1. The hat-snatching started as a fraternity prank, but quickly turned violent as people got rowdy and hat-wearers began to fight back. Eventually, the police and fire department had to be called in to subdue the rioters and a good number of them were hauled into court.

3 Trouser Skirts

Paris takes its fashion very, very seriously. So seriously, in fact, that wearing the wrong thing has actually caused a riot.

In 1911, two rival Parisian couture houses launched their "trouser skirts," an innovation in fashion that trod the very fixed line between the genders and seemed to promise greater flexibility for women in general. There were two different versions of the trouser skirt: One was a sort of baggy pants with a very low hanging crotch, described as "a sack with holes made for the legs to go through," not unlike the fashions on high streets today, and the other a pair of the same kind of pants topped with an over-skirt, again, not unlike high street fashions of today. Both versions were launched by models at the opening day of racing season to general revulsion and disgust, but thankfully, no violence.

In 1981 Miss New York became the first Miss America participant disqualified for "illegal use of padding" in the swimsuit competition.

It wasn't until the ladies attempted to promenade their future fashions on the boulevards that the fisticuffs started—at the Place de l'Opera, the poor models were attacked by a jeering mob of fashion Philistines, who pulled their hair, trampled their hats, and reduced them to tears. A squad of police officers on bicycles had to rescue the girls and escort them to safety.

4 Sheath Skirts

Riots in Paris we get—people in Paris love any excuse, good or not, to riot—but at anything-goes Coney Island? Bizarre, but true.

In 1908, two women clad in daring sheath, or Directoire, skirts—very tight, though long, skirts—were forced to take refuge in an automobile from an angry, pressing crowd of several hundred men and women. According to a contemporary report from the *New York Times*, the two women, attired in "steel gray" and "livid purple," respectively, walked in front of a restaurant with their dates. The couples were attempting to go to dinner when a crowd began to form around the women, "craning their necks and making remarks that did not please the wearers of the skirts." Police eventually came to the skirt wearers' rescue.

5 Any Clothes at All

In March 2009, a tourist was blamed for a "mini-riot" at a swinging sex party in an Australian nudist camp after he refused to remove his clothing. Really.

According to the owner of the White Cockatoo Resort in north Queensland, where the fracas occurred, the fight started when four female guests confronted one clothed man. The women complained that if he was going to see them naked, they ought to get to see him naked as well. The owner asked the man to remove his clothes, the man got angry, some "argy-bargy" (whatever that means) followed, the man was kicked off the premises, and the police were called.

When Maine prisons banned dental floss, an inmate sued, citing "stress and anxiety over the inability to fight tooth decay."

STAGGERING WORKS OF ARM-BREAKING GENIUS

SIX DANGEROUS WORKS OF ART

"Dangerous" art isn't always just edgy or provocative. Sometimes it's actually hazardous!

1 The Umbrellas

Starting in the 1960s married artists Christo and Jean-Claude began creating whimsical large-scale installations that redefined everything from Central Park to Parisian bridges. *The Umbrellas* was one of the pair's most ambitious projects. During the fall of 1991 the artists simultaneously installed 3,100 19-feet tall, metal and fabric umbrellas across California and Japan.

The piece was initially a triumph that drew three million visitors. Unfortunately, giant umbrellas are susceptible to high winds. A gust in California knocked down one of the nearly 500-pound umbrellas, killing 33-year-old Lori Keevil-Matthews. After Keevil-Matthews' death, Christo and Jean-Claude chose to end the exhibit early, only to have tragedy strike again when one of the workers dismantling the Japanese part of the work was electrocuted when the arm of the crane he was operating hit an overhead power line.

2 *Dreamspace V*

Maurice Agis' 1996 creation *Dreamspace V* was an 8,200-square foot inflatable network of translucent, polyurethane cells, large enough for people to walk through and explore. The interactive artwork toured the world for 10 trouble-free years, receiving thousands of visitors, until an unfortunate accident on July 23, 2006.

During an outdoor festival at Riverside Park in Chester-le-Street, England, a strong gust of wind from an approaching thunderstorm lifted the structure, snapping ropes that were meant to keep it tethered to the ground. One cell phone video and multiple closed-circuit cameras at the site recorded *Dreamspace V* as it thrashed around in the air. There were 30 people in the exhibit when it took off, and two visitors died as a result of their injuries. Agis was initially charged with manslaughter in connection with the deaths, but he was acquitted after a trial.

It's a Wonderful Life had an FBI file. In 1947, an analyst argued the film was an obvious attempt to discredit bankers.

Bodyspacemotionthings

When artist Richard Morris' *Bodyspacemotionthings*, London's Tate Modern's first interactive art experience, debuted in 1971, it allowed visitors to play on stand-up seesaws, swing on ropes, and roll around in giant concrete tubes. The work was a huge hit, welcoming 2,500 people in its first four days. However, those were also the exhibit's last four days.

The Tate had to close the exhibit because the visitors—as one guard memorably put it—"went bloody mad." The rambunctious crowd left the exhibit in shambles and suffered numerous minor injuries like splinters and bruises.

At least the injured patrons of the arts learned their lessons, right? Not quite. The Tate opened a modernized version of the exhibit in 2009, complete with design and construction materials that were meant to be safer and more resilient. Despite the safety upgrades, 23 people managed to injure themselves with everything from rope burns to head injuries to cuts and bruises in the exhibit's first week.

4-6 Three More Hazardous Tate Modern Offerings

Little did the Tate Modern know, but *Bodyspacemotionthings* accidentally set a precedent that transformed the museum into the world's most dangerous art gallery. Consider these three other works that tested patrons' mettle.

In 2006–2007, the Tate exhibited artist Carsten Höller's *Test Site*, a set of giant, spiraling slides that patrons could ride. Although no serious injuries were reported, the steep, twisting slides caused more than a few bumps and bruises.

Then, in late 2007, Colombian artist Doris Salcedo installed *Shibboleth*, a 500-foot-long crack in the floor. At its widest point, the crevice was three feet deep and one foot across. That might sound harmless enough, but three patrons fell in and received minor injuries; several others tripped just trying to walk around the piece.

Most recently, the Tate presented *Sunflower Seeds*, an installation by Chinese artist Ai Weiwei. The piece certainly didn't look dangerous; it featured more than 100 million ceramic sunflower seeds scattered on the floor. Ai intended the work to be interactive, hoping that viewers would walk across the artificial seeds, bury themselves in them, or lie down to make "seed angels."

Unfortunately, the thousands of patrons who tread on the work inadvertently kicked up ceramic dust, which turned out to be hard on the lungs. Curators were forced to rope off the installation just two days after it opened, fearing the dust would trigger asthma attacks.

WHEN PUSH COMES TO SHOVEL

EIGHT FAKE ARCHAEOLOGICAL FINDS

1 The James Ossuary

When this limestone coffin turned up in Israel in 2002, researchers thought it was the ossuary (container for skeletal remains) of Jesus' (yes, *the* Jesus!) brother James. On closer inspection, not quite. The ossuary itself dates back to the first century, but the carving on it that claimed that the remains were of the brother of Jesus is a modern forgery made to look old by the addition of a chalk solution.

2 The "Oldest" Star Map

In 1999 two metal detectors in Germany found a metal disc depicting the stars and planets. They claimed it was 3,600 years old and tried to sell it to German museums. A professor examined the disc and pronounced it no more than two or three hundred years old.

3 The Calaveras Skull

In February 1866, some miners in California found a human skull buried beneath a layer of lava. It fell into the hands of the State Geologist of California, who said the skull proved that humans, mastodons, and elephants had coexisted at some point in time in California. However, tests conducted at Harvard showed that the skull was of recent origin and one of the miners admitted the whole thing was a hoax.

4 Etruscan Terra-cotta Warriors

Even the experts can be duped on occasion. The poor Metropolitan Museum of Art got taken several times by the Ricardis, a family of art forgers. In 1915, the family sold a statue called *Old Warrior* to the Met. In 1916, they sold a work called the *Colossal Head*, which experts decided had been part of a seven-meter (about 23 feet) statue. The Met bought another piece, *The Big Warrior*, for $40,000 in 1918.

It wasn't until 1960 that tests of the pieces' glaze showed the presence of manganese, an ingredient that had never been used by the Etruscans. A sculptor who had been involved in the forgeries then came forward and signed a confession that the pieces were all fakes.

The first person arrested for speeding in the United States was going 12 mph. The year was 1899 and the cop was on a bicycle.

Piltdown Man

In 1912, diggers found pieces of a skull and a jawbone at Piltdown near Uckfield, East Sussex, England. Experts originally thought the bones belonged to an early form of man, but by 1953 scientists agreed that the specimen was actually the skull of a man with the jawbone of an orangutan.

6 The Cardiff Giant

Workers digging a well in Cardiff, N.Y., allegedly discovered this 10-foot-tall petrified "man" in 1869. In truth, sculptors had crafted the giant from a block of gypsum and shipped it to a farm in Cardiff, where it was buried for a year before being "discovered" by the well diggers. The giant then went on display; gawkers would pay 25 cents to gaze at it.

P. T. Barnum offered to buy the giant but the owners wouldn't sell. Barnum then had his own giant built; he claimed his was real and denounced the Cardiff Giant as a fake. On February 2, 1870, experts determined that both "giants" were fakes.

Michigan Relics

James Scotford "found" a slew of ancient relics, including a clay cup with symbols and carved tablets, in Michigan in 1890. Scotford and Michigan's Secretary of State, Daniel Soper, showed thousands of objects "found" in 16 counties of Michigan. The *Detroit News* even reported that the pair had found copies of Noah's diary. (Yes, the same Noah of ark-building fame.) Archaeologists agree that the artifacts were made with contemporary tools, and no more relics turned up after Scotford and Soper died.

8 Tiara of Saitaphernes

The Louvre's been on the wrong end of a hoax of its own. The museum purchased this regal artifact in 1896 on the assumption that it had belonged to Saitapherne, a third-century BCE Scythian king. Not so much, actually. It turned out that a skilled goldsmith had been commissioned to make the tiara for an archaeologist friend.

The embarrassed museum quickly stashed the piece away. Although it's no longer on permanent display, the Louvre still owns the tiara and breaks it out for the occasional exhibit on forgeries like the Salon of Fakes in 1954.

Why is "mayday!" an international distress signal? It comes from the French "*Venez m'aider*," or "Come help me!"

THE REVOLUTION WILL NOT BE TELEVISED (OR SUCCESSFUL)

FOUR CONSUMER REVOLUTIONS THAT DIDN'T DELIVER

The Amphicar: It's a Car! It's a Boat! It's a Huge Flop!

What It Promised: When West German company IWK rolled out its first Amphicars in 1961, it promised relief for anyone who was tired of having to own a separate car and boat. The hybrid vehicle's 43-horsepower Triumph motor promised to glide drivers across land or water in "the car that swims." It wasn't totally clear why anyone other than Batman or James Bond *needed* a car that doubled as a boat, but it still sounded cool! The Amphicar even came with a high-end price tag of $3,300.

What Went Wrong: Drivers/captains of the little convertible got a worst-of-both-worlds hybrid that would be familiar to anyone who's ever owned a futon. On land, the Amphicar was too slow; it needed more than 40 seconds to accelerate up to 60 miles an hour. In the water, it handled sluggishly since it didn't have a rudder. (Captains steered by turning the front tires, a system that wasn't super-responsive.) By 1968 IWK abandoned the brand.

More than 3,000 of the 3,700 Amphicars produced ended up in the United States, and hobbyists still roll them out on land and sea. The car-boats are actually hot commodities with collectors; one sold for $124,000 in a 2006 auction.

In 1955 Dodge introduced a car especially for women, a pink-and-white coupe with rosebud print interior called the La Femme. It flopped within two years.

Crystal Pepsi: Clearly a Debacle

What It Promised: In the early 1990s, clear products were all the rage. If something was clear, that meant it was pure, right? Even if the product in question was a super-processed cola. When Crystal Pepsi launched in 1992, it sought to offer drinkers a less sweet, caffeine-free cola experience that was so pure you could see through it.

What Went Wrong: Consumers saw through it, all right. Curiosity led to huge early sales in which Pepsi captured an additional one percent, or nearly $475 million, of the soft-drink market, but demand quickly dried up. Sure, the clear-cola novelty was interesting, but Crystal Pepsi wasn't very good. As Yum! Brands CEO David Novak, the man who had led the project, told *Fast Company* in 2007, "It would have been nice if I'd made sure the product tasted good." The drink that had caused such a splash in 1992 didn't even survive 1993.

3 The Internet: Where Nobody Is Saving Their e-Nickels

What It Promised: As e-commerce began to take off in the late 1990s, a number of companies like Beenz.com and Flooz.com set out to corner the market for online currency. Customers would buy special online credits that they could then redeem for merchandise in Web shops. Proponents claimed that these credits would fuel a whole new online economy.

What Went Wrong: Customers already had access to a pretty solid form of currency for their online purchases. They were called American dollars, and they were a bit more insulated from the vagaries of the dot-com boom and bust. Computer glitches played havoc with users' accounts—at one point Beenz.com accidentally gave $1,000 worth of its "beenz" to users—and the currency services became money-laundering hotbeds for credit card fraudsters. Despite massive funding—Beenz.com raised more than $80 million in investment, while Flooz.com had more than $35 million—the e-currency movement was dead by the end of 2001.

When Coca-Cola announced the return of Coke's original formula in 1985, ABC News interrupted *General Hospital* to break the story.

4 The Hulaburgers: McDonald's Says a Quick Aloha

What It Promised: McDonald's mogul Ray Kroc was facing an end-of-the-week problem in the early 1960s. Many of America's Catholics didn't eat meat on Fridays, which really put a dent in their desire to hit up Kroc's burger joints. Rather than offer a fish sandwich, Kroc decided to lure Catholics in with the Hulaburger, which was simply a slice of grilled pineapple between two slices of cheese on a toasted bun.

What Went Wrong: Kroc may have loved the Hulaburger, but customers weren't as crazy about it. In the days before the ubiquitous veggie burgers, diners were confused by a meatless burger. (Kroc later said customers chided him, "I love the hula, but where's the burger?") Luckily for McDonald's, another Catholic-friendly option fell into the chain's lap when Cincinnati franchisee Louis Groen invented a sandwich called the Filet-O-Fish. Kroc swallowed his pride—and possibly part of a Hulaburger—and pulled his failed brainchild.

Poor Kroc never totally gave up on his pineapple sandwich, though. He wrote in his memoir, "I still have one for lunch at home from time to time."

Staff members at New Jersey's Action Park were offered $100 to try the looping waterslide, but all refused after test dummies were dismembered.

10 LISTS OF LEMONADE

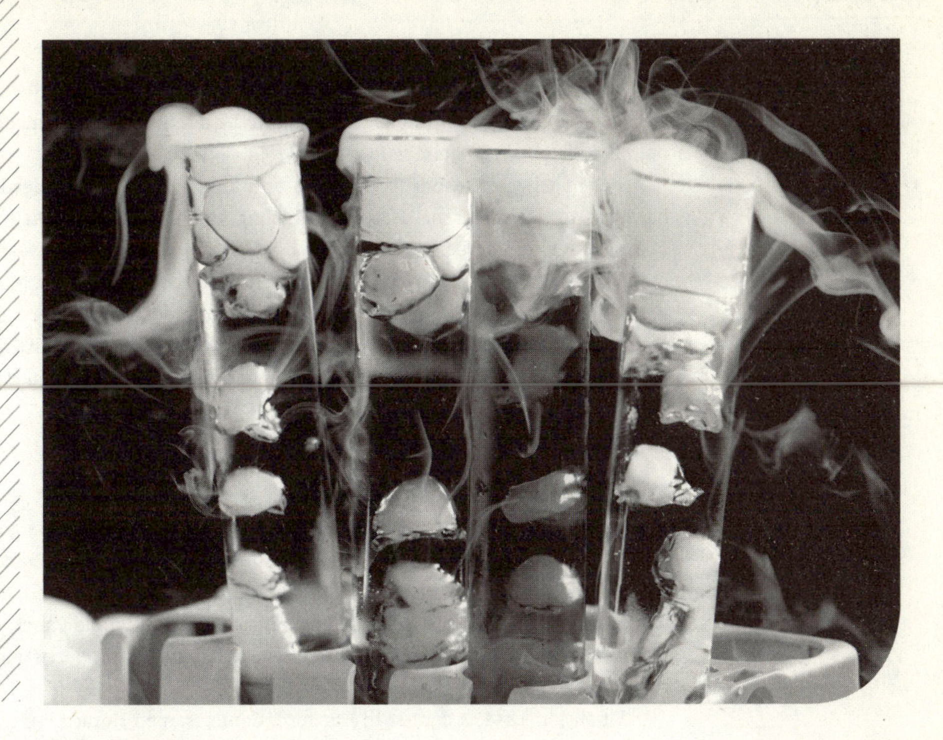

SPLENDIDLY BUNGLED

FOUR BRILLIANT SCIENTIFIC SCREWUPS

Penicillin (1928)

Mistake Leading to Discovery: Living like a pig
Lesson Learned: It helps to gripe to your friends about your job.

Scottish scientist Alexander Fleming had a, shall we say, relaxed attitude toward a clean working environment. His desk was often littered with small glass dishes—a fact that is fairly alarming considering that they were filled with bacteria cultures scraped from boils, abscesses, and infections. Fleming allowed the cultures to sit around for weeks, hoping something interesting would turn up, or perhaps that someone else would clear them away.

Finally one day, Fleming decided to clean the bacteria-filled dishes and dumped them into a tub of disinfectant. His discovery was about to be washed away when a friend happened to drop by the lab to chat with the scientist. During their discussion, Fleming griped good-naturedly about all the work he had to do and dramatized the point by grabbing the top dish in the tub, which was (fortunately) still above the surface of the water and cleaning agent.

On average, redheads require 20 percent more anesthesia than non-redheads. Scientists think the recessive gene that causes fair skin also affects the brain's pain receptors.

As he did, Fleming suddenly noticed a dab of fungus on one side of the dish, which had killed the bacteria nearby. The fungus turned out to be a rare strain of penicillium that had drifted onto the dish from an open window.

Fleming began testing the fungus and found that it killed deadly bacteria, yet was harmless to human tissue. However, Fleming was unable to produce it in any significant quantity and didn't believe it would be effective in treating disease. Consequently, he downplayed its potential in a paper he presented to the scientific community. Penicillin might have ended there as little more than a medical footnote, but luckily, a decade later, another team of scientists followed up on Fleming's lead. Using more sophisticated techniques, they were able to successfully produce one of the most lifesaving drugs in modern medicine.

2 Anesthesia (1844)

Mistake Leading to Discovery: Recreational drug use
Lesson Learned: Too much of a good thing can sometimes be, well, a good thing.

Nitrous oxide was discovered in 1772, but for decades the gas was considered no more than a party toy. People knew that inhaling a little of it would make you laugh (hence the name "laughing gas"), and that inhaling a little more of it would knock you unconscious. But for some reason, it hadn't occurred to anyone that such a property might be useful in, say, surgical operations.

Finally, in 1844, a dentist in Hartford, Conn., named Horace Wells came upon the idea after witnessing a nitrous mishap at a party. High on the gas, a friend of Wells fell and suffered a deep gash in his leg, but he didn't feel a thing. In fact, he didn't know he'd been seriously injured until someone pointed out the blood pooling at his feet.

To test his theory, Wells arranged an experiment with himself as the guinea pig. He knocked himself out by inhaling a large does of nitrous oxide, and then had a dentist extract a rotten tooth from his mouth. When Wells came to, his tooth had been pulled painlessly.

To share his discovery with the scientific world, he arranged to perform a similar demonstration with a willing patient in the amphitheater of the Massachusetts General Hospital. But things didn't exactly go as planned. Not yet knowing enough about the time it took for the gas to kick in, Wells pulled out the man's tooth a little prematurely, and the patient screamed in pain. Wells was disgraced and soon left the profession. Later, after being jailed while high on chloroform, he committed suicide. It wasn't until 1864 that the American Dental Association formally recognized him for his discovery.

Before John Madison Taylor invented his "reflex hammer" in 1888, doctors would simply whack patients' legs with their hands.

3 The Telephone (1876)

Mistake Leading to Discovery: Poor foreign language skills
Lesson Learned: A little German is better than none.

In the 1870s, engineers were working to find a way to send multiple messages over one telegraph wire at the same time. Intrigued by the challenge, Alexander Graham Bell began experimenting with possible solutions. After reading a book by Hermann Von Helmholtz, Bell got the idea to send sounds simultaneously over a wire instead. But as it turns out, Bell's German was a little rusty, and the author had mentioned nothing about the transmission of sound via wire. Too late for Bell though; the inspiration was there, and he had already set out to do it.

4 Iodine (1811)

Mistake Leading to Discovery: Industrial accident
Lesson Learned: Seaweed is worth its weight in salt.

In the early 19th century, Bernard Courtois was the toast of Paris. He had a factory that produced saltpeter (potassium nitrate), which was a key ingredient in ammunition, and thus a hot commodity in Napoleon's France. On top of that, Courtois had figured out how to fatten his profits and get his saltpeter potassium for next to nothing. He simply took it straight from the seaweed that washed up daily on the shores. All he had to do was collect it, burn it, and extract the potassium from the ashes.

One day, while his workers were cleaning the tanks used for extracting potassium, they accidentally used a stronger acid than usual. Before they could say "*sacre bleu!*," mysterious clouds billowed from the tank. When the smoke cleared, Courtois noticed dark crystals on all the surfaces that had come into contact with the fumes. When he had them analyzed, they turned out to be a previously unknown element, which he named iodine, after the Greek word for "violet." Iodine, plentiful in saltwater, is concentrated in seaweed. It was soon discovered that goiters, enlargements of the thyroid gland, were caused by a lack of iodine in the diet. So, in addition to its other uses, iodine is now routinely added to table salt.

Reed Hastings was inspired to start Netflix after racking up a $40 late fee on a VHS copy of *Apollo 13*.

THE MUMMY ON THE CAROUSEL

10 WEIRDEST PIECES OF UNCLAIMED BAGGAGE

What becomes of that one lonely suitcase lingering on the airport carousel? If it's lucky, it ends up at the Unclaimed Baggage Center in Scottsboro, Alabama, where you can find everything from the expected (clothing, toiletries, books) to, well, the stuff below.

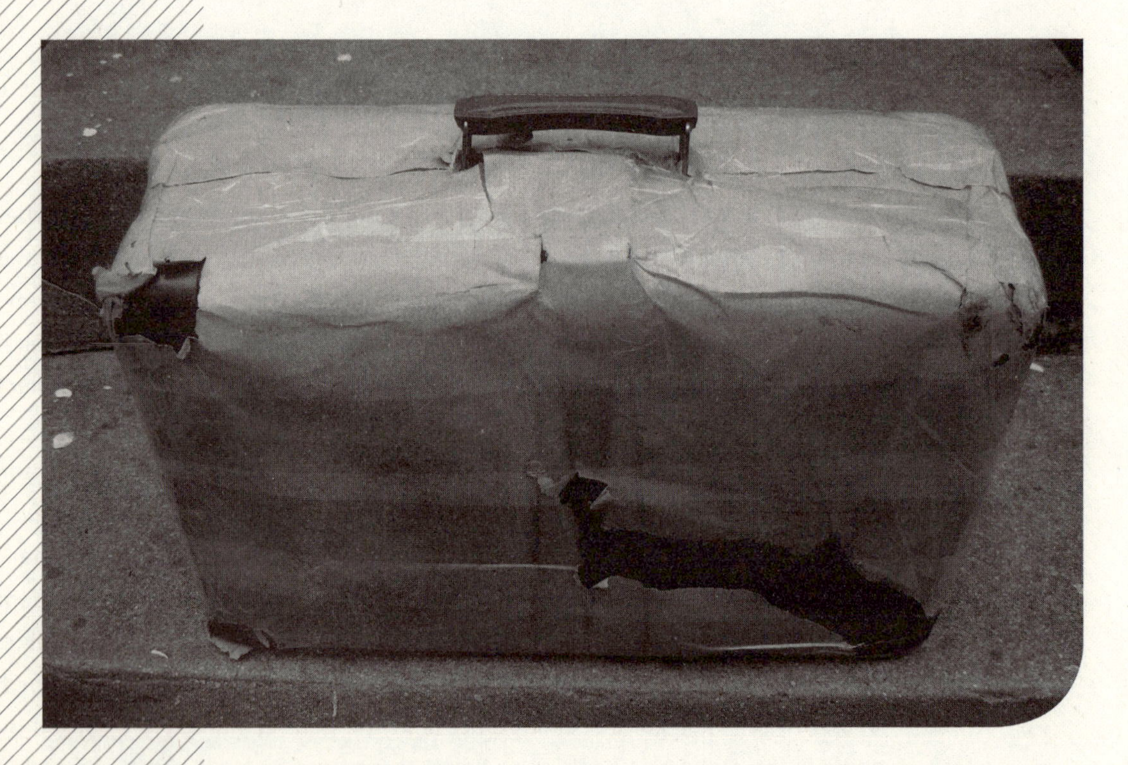

A slew of ancient Egyptian artifacts

From a mummified falcon to a shrunken head, a bunch of objects dating back to 1500 BCE turned up in an old Gucci suitcase. Christie's auction house ended up selling the museum-worthy items in the 1980s.

Frank Sinatra starred as a cardsharp in 1955's *The Man with the Golden Arm*—but his hands didn't. Milton Berle's hands served as stand-ins for some card-dealing close-ups.

Hoggle from *Labyrinth*

Unlike other items at the Unclaimed Baggage Center, Hoggle is not for sale. He is now a permanent part of the Unclaimed Baggage Center Museum. If you're not familiar, Hoggle was David Bowie's dwarf-goblin minion in the 1986 movie.

3. A rattlesnake—a live one

It was roaming free among the rest of the unclaimed baggage.

4. A naval guidance system

Yep, a piece of equipment worth $250,000 was lost and never claimed. The people at the Center decided to be good sports and return the expensive GPS to the Navy.

5. Rags-to-Riches Barbie

A woman purchased a Barbie for her daughter from the Center, which isn't at all unusual. After all, kids lose toys all the time. But when the girl yanked the head off her new doll, $500 in rolled bills tumbled out of her body.

6. A full suit of armor

Unlike the Egyptian artifacts, this guy was merely a replica of a 19th century piece. We're guessing that there's a guy out there who's still telling the story of the time he lost his suit of armor at the airport.

7. A violin from the 1770s

Like Hoggle, it resides in the Unclaimed Baggage Center Museum.

8. A 5.8-carat diamond set in a platinum ring

The owner had packed it in a sock.

9. A camera designed for use on a NASA Space Shuttle

Between this and the Navy's lost luggage, we're slightly concerned about the security measures our government agencies are taking! As with the Navy's guidance system, the Center dutifully returned the camera to NASA.

10. A 40.95-carat natural emerald

U.S. airlines made $5.7 billion from baggage fees and reservation change fees in 2010.

FIVE GREAT ESCAPES NOT INVOLVING STEVE McQUEEN (AND ONE INVOLVING A BEAR)

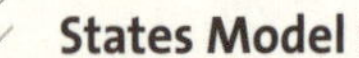

1 States Model School, Pretoria, Natal

This particular escape joins the ranks of history's greatest not because of elaborate planning or daring breakaways. Instead, its infamy is solidified by what might have later happened on the future stage of world history had it not occurred.

On November 15, 1899, Boer fighters in the colony of Natal (now part of South Africa) ambushed a British armored train, and, in the ensuing firefight, captured a British war correspondent for the *Morning Post*. To their delight, the new captive was the brave (some would say foolhardy) and adventurous son of a British lord—a catch with a great deal of leveraging power. But nearly a month later on December 12, the Boers' prize escaped the school in which he was imprisoned.

Bank robber Pretty Boy Floyd would occasionally destroy mortgage papers during heists, freeing hundreds of people from property debt.

Despite a price on his head, the plucky escapee successfully stowed away on board a train, slipped to safety into Portuguese East Africa (now Mozambique), and made international headlines. Within a year, the former prisoner of the Boers had returned to England and embarked on a political career that landed him a seat in Parliament. It also eventually got him a spot at 10 Downing Street, where he served as prime minister of Britain during World War II. His name, of course, was Winston Churchill.

2 Ghasr Prison, Tehran, Iran

On December 28, 1978, the Iranian government arrested Paul Chiapparone and Bill Gaylord, two executives of Texas-based Electronic Data Systems Corp., who were working overseas at the time. The Iranians accused EDS of trumped-up bribery charges and threw the two men into Tehran's notorious Ghasr Prison. Despite American concern for Chiapparone and Gaylord, the Iranian revolution against the Shah grew more and more chaotic, and the United States government seemed powerless to free them.

Enter EDS founder (and sometimes presidential candidate), H. Ross Perot. Taking matters into his own hands, Perot contacted retired Army Colonel Arthur "Bull" Simons, a Special Forces hero who had led a raid on the Son Tay POW camp during the Vietnam War, and asked him to rescue his two employees. Refusing payment for his services, Simons took a volunteer team of seven civilian EDS employees to Tehran, where they set about plotting a rescue. But before they could spring into action, Iranian revolutionaries stormed the prison. Hundreds of prisoners fled in the confusion, including Chiapparone and Gaylord. Shortly thereafter, they linked up with their would-be rescuers at Tehran's Hyatt Hotel and then escaped overland to Turkey with the aid of an Iranian EDS employee.

3 Yakutsk, Siberia

When Joseph Stalin's Red Army joined in Hitler's attack on Poland in 1939, the Russians bagged thousands of Polish soldiers as prisoners. Stalin promptly ordered hundreds of these men executed, while dispatching the rest to his brutal gulag labor camps in Siberia. Among those imprisoned was cavalry officer Slavomir Rawicz. While in Siberia, the resourceful Rawicz befriended the camp commissar's wife, and with her help, he and six other prisoners managed to escape during a blinding snowstorm.

A journey of epic proportions followed. A Polish teenage girl who had escaped her own camp joined Rawicz's band, and the ragtag group skirted Lake Baikal, slipped over into Mongolia, traversed the Gobi Desert, and crossed the Himalayas. After a journey of 4,000 miles, the Polish officer and his four fellow survivors staggered into British-controlled India, finally free. Amazingly, the irrepressible Rawicz soon returned to Europe to again battle the Germans. Rawicz's memoir, *The Long Walk*, continues to sell well, even though his amazing story has its doubters (some of whom point out its similarity to the Rudyard Kipling story "The Man Who Was").

Depression-era bank robber Willie Sutton was so pleasant and polite during his holdups that one victim described the experience as "like being at the movies."

Libby Prison, Richmond, Virginia

During the American Civil War, the Confederate government turned Richmond's three-storied Libby & Son Ship Chandlers & Grocers warehouse into what became known as Libby Prison. Not long thereafter, they crammed some 1,200 Union officers into it, many of whom, not surprisingly, spent their stays planning an escape.

But of all the break-out attempts at Libby, the most successful (and most elaborate) was masterminded by Colonel Thomas E. Rose. Using makeshift tools, he and a few fellow inmates tunneled down through a chimney, out of the prison's rat-infested cellar, underneath a vacant lot, and up into a shed some 50 feet away. Confident in his secretive route, the colonel returned to the prison on February 9, 1864, and led 15 other prisoners through the narrow tunnel and out into Richmond's unsuspecting streets.

Encouraged by Rose's success, 93 other prisoners quickly squirmed their way to freedom. Of those, an impressive 59 eventually returned to Union lines, making it the largest prison escape of the war. Although two officers drowned in the attempt, and the rest were recaptured, the Confederates couldn't help being impressed by their enemy's feat. The *Richmond Examiner* praised Rose's "scientific tunnel" and declared his breakout to be an "extraordinary escapade."

5 Alcatraz, San Francisco, California

Over the course of Alcatraz's three decades of operation as a federal prison in San Francisco Bay, it earned a fearsome reputation as America's escape-proof prison. But that didn't deter the dozens of inmates who attempted to flee "The Rock." Officially, none of these men succeeded, and at least seven died trying. But if any did pull it off, they were Frank Morris, John Anglin, and John's brother Clarence.

With the assistance of fellow inmate Allen West, the three men worked for months to carefully enlarge the vent holes in their cells to clear a way to the prison's roof. On June 11, 1962, after leaving carefully crafted dummy heads in their beds to fool prowling guards, they slipped out of their cells and, under the cover of darkness, reached the island's rocky shore. Relying on rafts made of their prison raincoats, Morris and the Anglin brothers entered the cold waters of San Francisco Bay to paddle for the mainland.

The following morning, Alcatraz guards discovered their absence and a massive manhunt ensued. But the three convicts were never recaptured. Although the FBI eventually concluded that they must have drowned, their bodies were never recovered. The resulting "what if" scenarios spawned the 1979 Clint Eastwood movie *Escape from Alcatraz* and, less directly, the annual Escape from Alcatraz triathlon in San Francisco. In the aftermath of the 1962 "escape" event, the federal government closed down Alcatraz the following year.

Al Capone's business card claimed that he was a used furniture dealer.

. . . AND ONE ADORABLE ANIMAL ESCAPE

Anyone who's watched a jailbreak movie knows that you won't get far without stealing a set of wheels. Even Juan knew that, and he was a bear. Juan, an Andean spectacled bear, made a daring escape from the Berlin Zoo in 2004. He rode a log across a moat designed to keep bears in their habitats, and then scaled a wall to gain his freedom.

Juan's first stop? The zoo's playground, where he terrified parents, rode the merry-go-round, and went down the slide. After a few minutes of play, though, he started to wander around again. Zookeepers needed a way to distract Juan, so they set a bicycle in his path. As Juan inspected the bike—possibly to see if it was a worthy vehicle for his ride to freedom—his handlers nailed him with a tranquilizer dart and carried the sleeping 300-pound bear back to his habitat.

When nine bison escaped from the Oakland Zoo in 1997, their keepers lured them back home using a trail of Wonder Bread.

GET RICH QUICK

FOUR PEOPLE WHO FOUND A FORTUNE

1 Lose a Hammer, Find a Hoard

In November 1992, a farmer living near the village of Hoxne in Suffolk, England, lost a hammer in one of his fields, so he asked Eric Lawes to use his metal detector to search for it. While looking for the hammer, Lawes happened upon something else of interest—24 bronze coins, 565 gold coins, 14,191 silver coins, plus hundreds of gold and silver spoons, jewelry, and statues, all dating back to the Roman Empire.

As required by British law, the men reported the so-called "Hoxne Hoard" to the local authorities, who declared it a "Treasure Trove," meaning it was legally government property. However, the same laws also require the government to pay fair market value for any trove. The farmer and Lawes split a cool £1.75 million, and the Hoxne Hoard went on permanent display at the British Museum.

Even the lost hammer got its 15 minutes of fame. It eventually turned up, and Lawes donated it to the British Museum.

2 The Declaration of (Financial) Independence

We've all heard of the man who bought a $4 painting at a garage sale, found an original copy of the Declaration of Independence inside, and sold the document for $2.4 million. A once-in-a-lifetime story, right? Not so much.

In 2006 Michael Sparks was visiting a Nashville thrift store, where he bought a candleholder, a set of salt and pepper shakers, and a yellowed print of the Declaration of Independence. Sparks figured the document was a worthless, modern reprint, so he paid the asking price—$2.48—and headed home.

After looking over the document for a few days, Sparks wondered if it might be older than he initially thought. With the help a few Internet searches, he realized he had purchased one of only 200 official copies of the Declaration of Independence commissioned by John Quincy Adams in 1820. Of those 200, 35 had been found intact; he had number 36.

Sparks spent a year having the print authenticated and preserved, and then he put it up for auction, netting a final sale price of $477,650.

The salt and pepper shakers, on the other hand, were still worthless.

Justin Timberlake's partially eaten French toast went to auction in 2000. It sold for $3,100.

3 The Art of Getting a Bargain

One day, an employee at a tool-and-die company in Indiana spent $30 for a few pieces of used furniture and an old painting of some flowers. When he got his new stuff home, he strategically hung the picture to cover a hole in the wall that had been bugging him.

Some years later he was playing a board game called *Masterpiece* in which players attempt to outbid one another for artwork at an auction. Much to his surprise, one of the cards in the game featured a painting of flowers that looked a lot like the one he had on his wall. A little research showed that his painting was very similar to the work of Martin Johnson Heade, an American still-life artist best known for landscapes and flower arrangements.

The Kennedy Galleries in Manhattan, which handles many of Heade's works, verified that the piece covering the hole in his wall was a previously unknown Heade painting, since named *Magnolias on Gold Velvet Cloth*. In 1999, The Museum of Fine Arts in Houston purchased the painting for $1.2 million dollars.

4 Arkansas is a Girl's Best Friend

When W. O. Basham found a 40.23-carat diamond in 1924 he wasn't deep in some South African mine. Nope, he was in Murfreesboro, Arkansas. The site where Basham made his find is perched atop a volcanic pipe (a geologic tube formed by an ancient underground volcanic explosion), which makes it prime diamond hunting real estate. The area is now Crater of Diamonds State Park, the only diamond site in the world that's open to the public. Best of all, the park's policy is: "You find it. You keep it. No matter how valuable it is."

Basham's big find—nicknamed "The Uncle Sam Diamond," the largest diamond ever discovered in North America—was later cut down to 12.42 carats. But his wasn't the last valuable rock dug out of that Arkansas soil.

In 1964 "The Star of Murfreesboro" emerged from the park to weigh in at 34.25 carats. Then, in 1975, came the 16.37-carat "Amarillo Starlight Diamond." The 6.35-carat "Roden Diamond" followed in 2006.

The crown jewel of the park, though, is a comparatively small 3.03-carat rock. The "Strawn-Wagner Diamond turned up in 1990 and was later expertly cut down to 1.90 carats. Despite its smaller size, the Strawn-Wagner stands out because it earned a "Perfect" rating by the American Gem Society—the first diamond to ever receive such a high grade.

Don't think this list of big gems means the site has been tapped out. On average, searchers find two rocks every day at Crater of Diamonds. They're not all as big as the Uncle Sam Diamond, but maybe you'll get lucky. There's only one way to find out . . .

British antiques dealer Morace Park bought a film canister on eBay for around $5 in 2009. The can contained something more valuable: an unreleased Charlie Chaplin short called *Zepped*.

ATTENTION: GOLD DIGGERS!

FOUR LOST TREASURES JUST WAITING TO BE FOUND

Sure, you missed the boat on those four instant fortunes, but don't despair! Get busy trying to find one of these.

1 The Lying Dutchman?

Arthur Flegenheimer, who went by the alias "Dutch Schultz," was a New York mobster during the 1920s and 1930s known for his brutality and hard-nosed business tactics. By the time he was 33, Dutch had taken on the Mafia in numerous gangland wars, fought the U.S. government twice on tax evasion charges, and amassed a fortune.

As his second tax evasion trial took a turn for the worse, it appeared Schultz was looking at jail time. In preparation, he placed $7 million inside a safe, drove to upstate New York, and buried it so he'd have a nest egg when he got out of prison. The only other person who knew the location of the safe was the bodyguard who helped him dig the hole. Shortly after burying the booty, both mobsters were gunned down by hitmen in New Jersey.

On his deathbed, Schultz began hallucinating and rambling. A court stenographer came in to record his statements and some believe his incoherent references to something hidden in the woods in Phoenicia, New York, might be a clue to the location of his buried loot. Of course the meaning of his words is cryptic and not 100 percent reliable, but that hasn't stopped hundreds of people from looking. So far, though, Dutch's safe has not been found.

2 Straight Out of the Poe House

Before Edgar Allan Poe was Edgar Allan Poe, he was just another struggling writer who couldn't catch a break. In 1827, he hired Calvin F. S. Thomas to publish 50 copies of his manuscript, *Tamerlane and Other Poems*, in the hopes that it would kick-start his career. Unfortunately, *Tamerlane* received no critical consideration at the time (and has only received middling reviews since), so Poe's rise to fame would have to wait until he published *The Raven* nearly 20 years later.

Because the book had such a small print run, first editions have become some of the most sought after pieces in American literature. Only 12 copies are known to still exist, mostly held by libraries and museums. But there could easily be more that have gone unnoticed because, for reasons unknown, Poe's name does not appear as the author of the book; it is only attributed to "A Bostonian." Without a familiar name on the cover, many people dismiss

"All the gold in Fort Knox" is worth roughly $125 billion.

Tamerlane as a worthless collection of poems by some anonymous writer no one's ever heard of. The last copy to turn up went for a mere $15 at an antique store in 1988. At auction a month later, the book fetched $198,000.

3 A 10-Cent Treasure

Think a dime is nearly worthless? Think again. A wagon train left Denver in 1907 carrying six large barrels filled with newly minted "Barber" dimes, nicknamed after Charles Barber, the designer of the coin. The dimes were being delivered to Phoenix, Ariz., some 900 miles away, but the shipment never arrived. One theory is that bandits attacked the wagon train. Others believe the party might have plummeted hundreds of feet to the bottom of Colorado's Black Canyon while navigating the treacherous mountain trails. All that can be said for sure is that neither the coins, nor the men carrying them, were ever seen again.

Now, a little over 100 years later, a single 1907 Barber dime in excellent condition fetches around $600. Assuming the barrels weren't destroyed and the coins haven't been exposed to the elements all this time, these missing coins should be fairly flawless. If you estimate 5,000 coins at $600 each, you're looking at $3,000,000. With that kind of dough, you could make an awful lot of phone calls.

4 Morriss' Code

In 1820, a mysterious stranger left a locked iron box with Robert Morriss, an innkeeper in Bedford County, Virginia. The stranger, who went by the name Thomas Jefferson Beale, said that a man would be coming to retrieve the box some time in the next 10 years. However, if no one ever came, Morriss could keep the box and the contents inside.

What was inside the box? Beale reluctantly revealed that there were three pages covered in numbers. These "ciphertexts" were coded messages that could only be read by using corresponding documents as a key. Beale promised to send the three keys to Morriss when he arrived in St. Louis, so that, should the box become Morriss', he could decipher the messages and learn the location of a treasure Beale had buried nearby.

Twenty years later, no one had ever come for the box, nor had Morriss received any key documents from St. Louis. He went ahead and opened the box, and spent the rest of his life trying to decode the pages to no avail. After his death, Morriss left the box to a friend, who, surprisingly, was able to decipher the second page using a particular copy of the Declaration of Independence. The page described the treasure itself—2,900 pounds of gold, 5,100 pounds of silver, and thousands of dollars' worth of jewelry. The message then went on to say that the exact location of the treasure was found on the first page, so you would have to decode it to find the loot. The first and third pages have never been deciphered, despite people working on it for nearly 175 years.

One of the first known items made of aluminum was a baby rattle designed for Napoleon III's son in the early 1800s. At the time aluminum was so rare it was more precious than gold.

REPETITION IS KEY

FIVE FAMOUS STUTTERERS

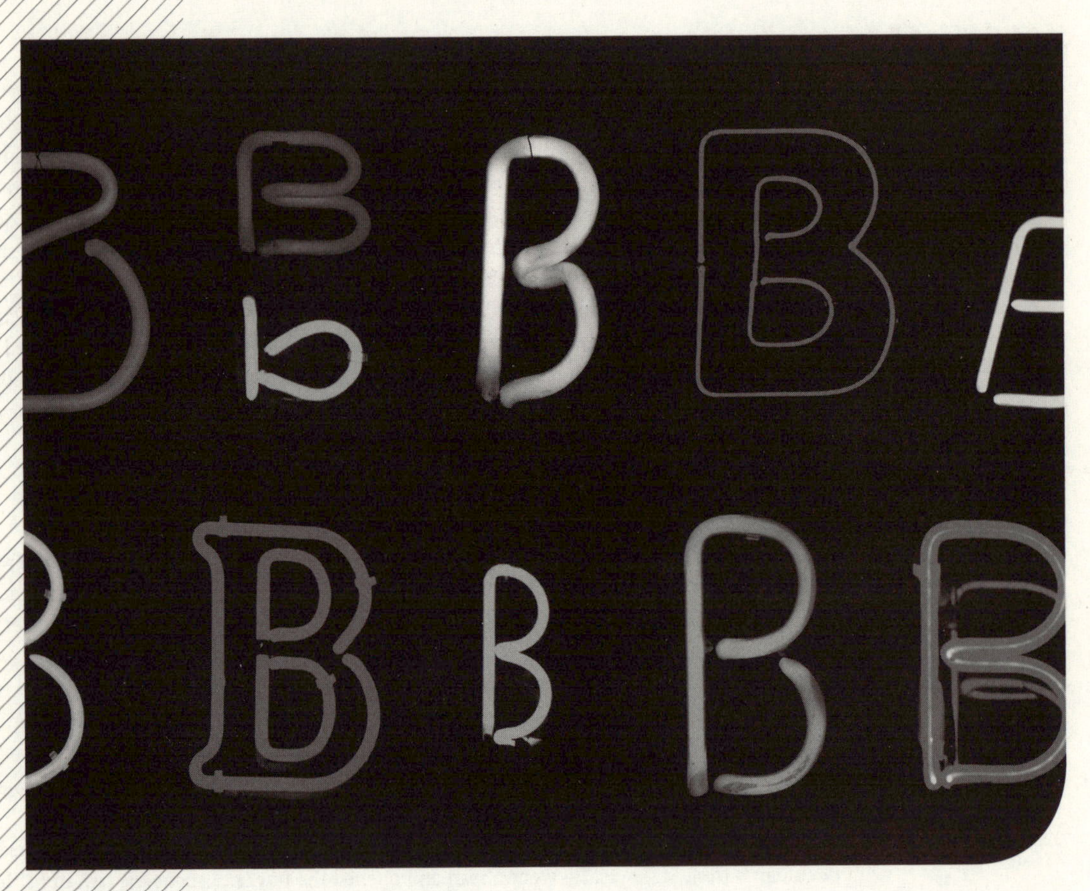

Joe Biden

A keynote speaker at a National Stuttering Association convention in 2004, Biden grew up with a stutter and was teased by his Catholic schoolteacher. Apparently, the nun called him "B-B-B-Biden" in front of the whole class. When young Joe told his mom what happened, she stormed back to school and chewed out the sister. Biden practiced reading aloud in front of a mirror and, later, in law school, became friends with another stutterer with whom he practiced his speech.

In 1939 Utah-based Loftus Novelty began producing rubber chickens en masse. Today it's still the only American manufacturer of rubber chickens.

Moses

We admittedly can't say if Moses stuttered or not, but if you believe the Good Book, here are some lines that indicate he most certainly may have:

Exodus 4:10

. . . I am slow of speech and of a slow tongue . . .

Exodus 4:14-15

And the anger of the Lord burned against Moses and He said, "Is not Aaron the Levite thy brother? I know that he can speak well . . . thou shalt speak to him, and put the words in his mouth . . . "

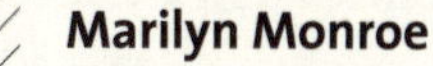

Marilyn Monroe

Several books on stuttering and stutterers, suggest that Marilyn's breathy, sensual delivery was actually a way of dealing with a stammer. Apparently a speech coach instructed Monroe to use exaggerated mouth movements to keep her mild stutter in check.

4 James Earl Jones

Hard to imagine Darth Vader stuttering, isn't it? Some actors who stutter are able to curb their speech impediments while on script but continue stuttering in everyday conversation, a condition called situational fluency. Such is reportedly the case with one of Hollywood's most famous voices, James Earl Jones. If you saw him in the movie *A Family Thing*, then you know he can pull off a believable stutter. That's because he really does. The story goes like this: Jones accidentally stammered while reading lines to the film's director Dick Pearce. Pearce thought the trip-ups made the character seem more vulnerable and asked Jones to keep doing it.

Lewis Carroll

Many people know Lewis Carroll's real name was Charles Lutwidge Dodgson. But how many people know that this son of a clergyman wanted to become a priest? Apparently he was unable because his stutter would have made it very hard come sermon time.

Marilyn Monroe's ghost allegedly still wanders around her suite at Hollywood's Roosevelt Hotel, where she reportedly admires her reflection in a mirror.

THAT'S MY SLICE OF THE COLOR WHEEL!

10 TRADEMARKED COLORS

Have you mixed up a brilliant color for your business but worry that a competitor might swipe it? Have no fear! You can actually trademark a tiny bit of the rainbow for your business, just like these 10 groups did.

1. Qualitex Green-Gold: You may not know this color, but *Qualitex v. Jacobson Products Co., Inc.*, is what put "colormarking" on the map. Qualitex used a unique blend of green and gold on its dry cleaning presses, but in 1989 competitor Jacobson began using the exact same shade, allegedly to confuse businesses into buying their product instead. Qualitex successfully sued Jacobson, and the practice of "colormarking" was born.

2. Tiffany Blue

3. Target Red

4. Cadbury Purple

5. Barbie Pink

6. Home Depot Orange

7. T-Mobile Magenta

8. Wiffle-Ball Yellow

9. UPS Brown

10. Coca-Cola Red

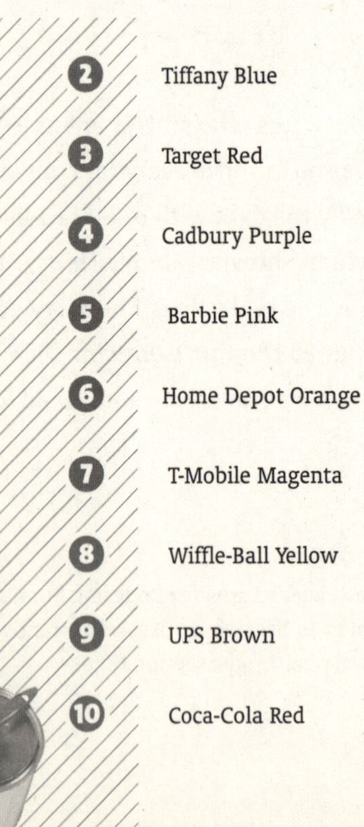

British merchant Peter Durand patented the tin can in 1810. Ezra Warner patented a can opener in 1858. In between, people opened cans with chisels and hammers.

DROPPING KNOWLEDGE

SIX CREATIVE USES FOR POOP

A wise book once told us that everyone poops. Not everyone knows just how useful scat can be, though.

Paper

Need a sheet of paper? If you've got some time and some elephant dung, you can make your own. Poop is composed largely of indigestible fiber, and in Thailand and China this fiber is being turned into stationery. Artisans collect elephant patties, which they then clean, spin, dye, and dry to make paper. The proceeds from the sale of the paper support elephant preservation projects and sanctuaries.

2 Medicine

When the German army occupied northern Africa during World War II, many soldiers suffered terribly from dysentery. Local residents seemed to take the illness in stride, though. What was the locals' secret? It definitely wasn't for the weak of stomach. As soon as a local noticed the first symptoms of dysentery, they followed a camel. When the camel pooped, the dysentery patient ate the droppings.

The first products designed specifically to wipe one's nethers were aloe-infused sheets of manila hemp dispensed from Kleenex-like boxes introduced in 1857.

Gross? Totally. Effective? You bet, as long as the droppings were fresh. The camel dung was full of a beneficial bacteria, *Bacillus subtilis*, that aided in digestions and crowded out the dysentery bacteria.

The fastidious German troops were horrified to learn what the cure entailed, but it beat having dysentery. Eventually the German medical corps managed to isolate the bacteria to make the treatment infinitely more pleasant.

3 Camouflage

With magical dysentery-curing powers like those, it's easy to see how camel "apples" became lucky charms for German troops in World War II. At least they were until the Allies discovered the German habit of intentionally running tanks over piles of the droppings for good luck. The Allies then developed and planted land mines that looked like camel dung!

When the Germans caught on to the trick, they began avoiding fresh piles of camel manure. In turn, the Allies began making mines that looked like camel dung *that had already been run over by a tank* and therefore seemed safe enough to a Nazi driver. Brilliant!

4 Coffee

The world's priciest coffee comes from poop. The Asian palm civet is a weasel-like little critter that loves to eat coffee cherries on the Asian regions where coffee is grown. The civets can only partially digest the cherries, though, so the coffee comes out the other end largely intact. The excreted beans can then be gathered and washed, and the coffee beans are sold as "kopi luwak," which can cost hundreds of dollars per pound. The partial digestion process adds a wonderful flavor and complexity to the coffee.

5 Gunpowder

You might know that "guano" is the polite term for the droppings of seabirds, bats, and seals. Guano's usefulness isn't as well advertised, though. Guano is full of nitrogen, particularly potassium nitrate, which can be used to make gunpowder. Its high phosphorous levels also make guano a dynamite fertilizer.

6 Clean Water

Years of tin and silver mining in Bolivia have really hurt the country's water quality. Acidic water laden with the metals leach out of the mines, and the toxic runoff pollutes water, killing algae and fish. What to do? One method of cleaning the water involves introducing bacteria that processes the harmful sulfates in the water by binding the dissolved metals into compounds like iron sulfide and zinc sulfide so that the metals sink to the bottom of the water.

Where does the water-cleaning bacteria originate? In llama poop. Scientists add llama dung to water treatment lagoons and wetlands, and the bacteria get to work scrubbing out those nasty sulfates.

New York, Chicago, and Boston all have feral parrot populations.

LET YOUR MIND WANDER

THE UNLIKELY INSPIRATION FOR FIVE BRILLIANT INVENTIONS

Boring Sermon a Godsend for Note Takers

Arthur Fry had two big problems during a 1974 church service. First, the scraps of paper he kept using as bookmarks in his hymnal kept falling out. Second, the sermon was incredibly dull. As Fry's mind wandered during the boring sermon, he realized what he needed was a sticky, reusable bookmark that he could slap on his hymnal's page without damaging them.

Fry, a scientist at 3M, remembered hearing about a weak adhesive that had been kicking around 3M's shelves without any real commercial use. The next day he set to work putting the adhesive on small pieces of paper, and the Post-it Note was born!

Albert Einstein helped patent several refrigerator designs, though none of them proved successful.

Ruined Candy Bar Paves the Way for Frozen Burritos

In 1945 radar engineer Percy Spencer had a pretty rough day. Spencer was hard at work at Raytheon when he stood in front of a magnetron, a high-powered vacuum tube that generates microwaves. After a few moments, he realized that the candy bar in his pocket was melting.

The melted candy bar was rough news for Spencer's pants pocket, but it was a major coup for future lovers of Hot Pockets. Spencer eventually harnessed the magnetron's power to cook food, and the world suddenly had the microwave oven for all of its quick-warming needs.

Dirty Hands Make Sweet Breakthrough

As a general rule, you should probably wash your hands pretty thoroughly after a long day of working in a chemical lab with coal-tar derivatives. In 1879 Constantin Fahlberg was eating dinner after getting off of work at Johns Hopkins University when he noticed all of his food tasted supersweet. Although he later claimed he had "washed [his] hands thoroughly"—whatever you say, Constantin!—Fahlberg realized that some chemical he'd been working with that day was an ultra-potent sweetener. Turns out that Fahlberg had accidentally discovered the calorie-free sweetener saccharin. Score a rare victory for halfhearted hand washing!

4 Brilliant Invention Sticks to Dirty Dog

In 1941, Swiss engineer Georges de Mestral went for a walk in the woods with his dog. When de Mestral and his pooch returned home, they faced a familiar problem to anyone who's been on a rural ramble: burrs had clung to their clothes and fur.

Most of us just get irritated and pick the burrs off, but de Mestral turned curious. He decided to take a peak at a burr under a microscope and noticed it was covered with tiny hooks that clung to anything that passed by. De Mestral realized that he could craft a fastener that involved similarly tiny hooks grabbing onto little loops, and after years of tinkering he perfected Velcro.

5 Ugly Carrots Can Have Babies

The next time you enjoy baby carrots at lunch, think of the ugly carrots that inspired California farmer Mike Yurosek. From the time he started farming carrots in the 1960s, misshapen carrots plagued him. He couldn't sell the unappealing veggies to supermarkets, but tossing them out seemed incredibly wasteful. Finally, in 1986, Yurosek had an epiphany: every ugly carrot had a picture-perfect carrot inside it just waiting to get out. Using industrial peelers and slicers, Yurosek began whittling these misfit veggies into photogenic "baby carrots."

Yurosek was cautiously optimistic when he sent his first batch of baby carrots to a supermarket. Even he couldn't have anticipated the tiny little guys' success, though; the supermarket quickly called and told Yurosek that from now on they wanted only baby carrots. By the time Yurosek died in 2005, nearly half a million tons of baby carrots were making their way into lunch boxes each year.

Duct tape debuted in 1942 for use by the U.S. Army as a waterproof sealing tape for ammunition boxes.

YOU'VE MADE IT THIS FAR!

FIVE DOOMSDAYS WE'VE ALREADY SURVIVED

Some Fuzzy Papal Math, 1284

When the Pope decrees something, people tend to listen. And they did in 1213 when Pope Innocent III wrote that the end of the world was coming in 1284. How did Innocent III arrive at that date? He added the 666 years mentioned in the Book of Revelation to the date of the founding of Islam.

We're Going to Need a Smaller Boat, February 20, 1524

German scholar Johannes Stöffler was better at math and astronomy than he was at predicting the apocalypse. Stöffler made calculations that indicated a flood of epic proportions was going to engulf Earth on February 20, 1524. One German nobleman, Count von Iggleheim, took the prediction so seriously that he borrowed a trick from Noah and built his own ark.

Comedian Carrot Top's standard contract politely asks that venues not serve him carrot cake because "It's still not funny."

When a light rain did begin falling on the appointed day, people panicked and began a mad rush to board von Iggleheim's ark. Hundreds died as they attempted to fight their way onto the boat, and the Count ended up being stoned to death.

3 Sinners in the Hands of a Lousy Predictor, 1697 and Others

Famed Puritan minister Cotton Mather didn't have the apocalypse-predicting acumen to match his fiery sermons. Mather's first analysis of Revelation revealed that the world would end in 1697. When the year ended without so much as a minor doomsday, Mather revised his calculations and arrived at some pleasant news: we all had until 1716 to live. When that date didn't pan out, either, Mather made his final revision: the world was ending in 1724. Seriously this time, guys.

4 Run! Halley's Comet Is Tailing Us!, 1910

It's a good thing Halley's Comet is only visible every 75 years or so, because people have a tendency to freak out when the comet is near. The 1910 passage of the comet led to some particularly outlandish behavior when some scientists became worried that the comet's tail would pass through our atmosphere and fill it with poisonous gasses that would kill us all. A hysteria ensued, and people spent fortunes on gas masks, anti-comet pills, and comet-proofing for their homes. The comet passed uneventfully on the night of May 18, 1910. Turns out the gas in a comet's tail is so diffuse that it's not really much of a threat.

5 Earth and Jupiter Get in Line, March 10, 1982

The Halley's Comet scare of 1910 may have been a bust, but it didn't stop doomsayers from making bold astronomical predictions. In 1974 respected scientists John Gribbin and Stephen Plagemann laid some pretty heavy news on the world in their book *The Jupiter Effect*. In March 1982 several planets would be in alignment on one side of the sun. According to the authors' theory, the alignment would basically short-circuit nature and cause a flurry of massive earthquakes. Los Angeles fared particularly poorly in their prediction; action on the San Andreas fault was poised to decimate the city.

When the fateful day came and went without so much as a tremor, Gribbin and Plagemann had some explaining to do. Why not sell a book or two while they were at it? In 1983 the pair published the follow-up *The Jupiter Effect Reconsidered*, although Gribbin would later renounce the unproven "effect" entirely.

Just before the Nazis invaded Paris, H. A. and Margret Rey fled on bicycles. They carried their manuscript for *Curious George*.

10 LISTS THAT DIDN'T FIT NICELY INTO ANY OTHER CHAPTER

GOING POSTAL

FIVE STAMP STORIES WORTH RETELLING

1 The Stamp That Divided a Nation

Never underestimate the political power of the stamp. When the American Civil War broke out in 1861, the seceding Confederate states snatched up a good bit of government property. This included everything from forts to arsenals to thousands of post offices stocked full of stamps. Not wanting the enemy to profit off their goods, the Union recalled every U.S. stamp ever issued and declared them invalid for postage. Instead, people were allowed to exchange their old stamps for replacements, which the government had quickly printed with new designs.

2 The Stamp with All the Right Intentions, and All the Wrong Music

In 1956, East Germany decided to honor the death of native composer Robert Schumann by featuring him on a stamp. The design included a commemorative portrait of the artist against the backdrop of one of his musical scores. All well and good, except the musical manuscript they used was that of fellow composer Franz Schubert. Close, but no cigar. The stamps were recalled and replaced with ones showing music actually written by Schumann.

3 The Stamps Made from Stolen Maps

During World War I, the Baltic region of Latvia didn't have much to call its own. It was governed by Russia, and German forces occupied much of the area. In 1918, however, Latvia gained independence during the chaos and collapse of the Romanov dynasty. In addition, German forces had retreated . . . but not without leaving their mark on the new nation. Oddly enough, that mark was on Latvia's stamps.

Latvia suffered devastating damage during the war. Factories were destroyed or moved to Russia, and paper was in short supply. So when the young nation got ready to print its first national stamps, postal officials got creative and used the blank backs of German military maps and unfinished banknotes. Indeed, if you look on the underside of some Latvian stamps from this era, you'll see a tiny sliver of a military map used by the Germans during World War I.

4 The Stamp That Almost Started a War

Don't be fooled by its size. A tiny little stamp can cause big trouble. Case in point: This stamp issued by Nicaragua in 1937. Not uncommon, the stamp featured a map of the country, but it included a large section of land also claimed by neighboring Honduras. The two countries

The glue used on Israeli postage stamps is kosher.

had long disputed ownership of the region, and it remained a source of great contention. In 1906, King Alfonso XIII of Spain decided the matter in favor of Honduras, but Nicaragua refused to acknowledge the decision.

Tensions grew in the following years, so when Nicaragua released the stamp in 1937, Hondurans were outraged. Government officials, newspapers, and radio stations demanded the stamps be recalled and destroyed. Nicaraguan authorities, however, refused and insisted the map was correct. They also pointed out that they had the courtesy to label the area on the stamp as *territorio en litigio*. Regardless, in a matter of weeks, anti-Nicaraguan demonstrations erupted in the Honduran capital of Tegucigalpa. Across the border, Nicaraguan radio announcers called for military action, demanding the national army be sent to guard the border region. The public even began a donation drive designed to fund more planes to build up the Nicaraguan Air Force.

At the last minute, the United States, Costa Rica, and Venezuela intervened to defuse the conflict before it escalated into war. Both countries agreed to withdraw their armed forces from the disputed area and stop mobilizing troops. And, naturally, the peace agreement called for withdrawing the offending stamps. They evidently remained in circulation, however, until supplies in private hands ran out.

5 The Stamp That Moved the Panama Canal

In 1902, the U.S. Congress was poised to pass legislation to link the Pacific Ocean and Caribbean Sea with a canal across Nicaragua. That is, until engineer Philipe Bunau-Varilla (and a certain stamp) got involved.

In the 1880s, Bunau-Varilla worked for a French company that had attempted to construct a similar canal across Panama. But engineering difficulties, financial mismanagement, and deadly yellow fever epidemics eventually bankrupted the company and prevented it from completing the project. Still believing Panama (then part of Colombia) presented the best route for such a canal (and still wanting a government contract to construct it), Bunau-Varilla lobbied Congress to switch its plans, claiming Nicaragua's terrain was too unwieldy. Then, in the spring of 1902, nature worked in his favor. Mt. Momotombo, a volcano in Nicaragua, erupted.

Knowing the incident would sway the American canal vote, Nicaraguan officials immediately began denying reports of the eruption, and Bunau-Varilla was left struggling for a way to counter the Nicaraguan cover-up. Fortunately, he remembered once seeing a Nicaraguan postage stamp featuring Mt. Momotombo, conveniently depicted with smoke rising from the top. After rummaging through stamp shops in Washington, he found the one he was looking for and promptly purchased 90 copies. In a matter of days, all 45 U.S. senators had received the Mt. Momotombo stamp, complete with Bunau-Varilla's caption, "An official witness to volcanic activity in Nicaragua." This menacing volcano, they were told, would threaten the canal route. Sure enough, when the Senate voted on June 19, 1902, the Panama route won.

Ethel Merman recorded a song to the tune of "Zip-a-dee-doo-dah" to promote zip codes when they were first introduced.

TROOP LOOKS LIKE A LADY

FOUR MEMORABLE MOMENTS IN CROSS-DRESSING HISTORY

1 Cross-Dressing to Join the Army

Until recently, women have rarely been allowed to serve as soldiers. So what was a gal to do if she wanted to serve her country? Naturally, disguise herself as a man and join the troops. At least 400 Civil War soldiers were women in drag. These included Union Army soldier "Frank Thompson" (also known as Sarah Edmonds), whose small frame and feminine mannerisms (rather than causing suspicion) made her an ideal spy, as she could spy on the Confederates disguised as . . . a woman!

She wasn't the first woman to don a male disguise and join the army, though. During the Revolutionary War, women fought as men on both sides. Hannah Snell, for example, joined the British army to find her husband, who had walked out on her to enlist. Once her true sex was discovered (thanks to a pesky groin injury), she became a national celebrity in Britain, and made a postwar career of performing in bars as the "Female Warrior."

Kathryn Johnston was the first girl to play Little League. In 1950, she cut her hair, called herself Tubby, and made a team in Corning, N.Y.

❷ Cross-Dressing to Keep a Royal Family Together

With all the power struggles that went on in the court, the French royal family would go to great lengths to avoid sibling rivalry. In one of the more extreme cases, Philippe I, Duke of Orleans (1640–1701), was raised as a girl to discourage him from any political or military aspirations. This would make things easier for his brother, the future King Louis XIV. Philippe wore dresses and makeup, enjoyed traditionally feminine pursuits, and was even encouraged toward homosexuality.

A girlie man he might have been, but he married twice and even had a mistress. When necessary, he could even lead an army into battle. (This is the nation, after all, that gave us that famous cross-dresser Joan of Arc.) A brave commander, he would go into battle wearing high heels, plenty of jewelry and a long, perfumed wig. One of his wives claimed that Philippe's biggest fear when going into battle was not bullets, but the possibility of looking a mess. He avoided gunpowder (with the black smoke stains) and didn't wear a hat, to avoid ruining his hair.

❸ Cross-Dressing to Commit Espionage

There have been many instances of cross-dressing spies, but one of the most impressive deceptions in history was carried out by Shi Pei-Pu, a singer with the Beijing Opera (in which, traditionally, all roles are played by men). In 1964 he disguised himself as a woman to seduce Bernard Boursicot, an attaché in the French Foreign Service. Their affair lasted 20 years (on and off), during which Boursicot passed several official documents to Shi, believing that "her" safety was at risk if he didn't participate. After they were separated in 1965, Shi came back into Boursicot's life by claiming to be pregnant, and even revealed a baby boy. They later lived as a family. The happy couple was eventually arrested for espionage in 1983, and Shi's secret was revealed, *Crying Game* style, to the stunned Boursicot.

❹ Cross-Dressing to Get Rowdy

Hindu women in India have traditionally lived inhibited lives, tending the home for their families. But on one night each year, in the city of Jodhpur, they come to life at the so-called Festival of Fun. Dressed as noblemen, complete with turbans and large fake mustaches, they walk the streets in gangs, brandishing sticks, beating any males who are foolish enough to be out there.

The festival celebrates an ancient domestic dispute between the Hindu god Shiva and his wife Ganwar. While the women sing devotional songs asking the goddess to return to her husband, they also take this as their only opportunity to do what they long to do all year: behave like men. Their husbands, respecting the tradition, let them go wild. It sounds like fun (as long as you're not a clueless male who forgot to stay at home that night), but one thing concerns me: If a woman's disguise is especially good, does she risk being mistaken for a man and beaten up by her friends?

The famous leg featured in posters for *The Graduate* belongs not to Anne Bancroft, but to a then-unknown Linda Gray (later of *Dallas*).

JOLLY ROGERS AND REGINAS

FIVE LITTLE-KNOWN PIRATE STORIES

Yo-Ho-Ho and an Epidural

Grace O'Malley was the Irish Sea Queen of the 16th century. Earning her sea legs as a kid on voyages with her father, O'Malley went on to lead a crew of 200 sailors as part of her Celtic Sea "protection service." Her specialty? Intercepting merchant ships to negotiate their safe passage to Galway and ruthlessly pillaging any "uninterested customers."

Infamous for being lewd, gambling too much, and cussing like—well—a sailor, O'Malley truly proved her mettle when she gave birth mid-voyage. Soon after the delivery, Turkish pirates attacked the ship, and when the flailing crew came running to O'Malley, she reportedly snapped, "May you be seven times worse off this day 12 months from now, you who cannot do without me for one day!" When the postpartum hell-raiser finally emerged on deck waving her gun, the attackers quickly remembered they had other engagements.

Papua New Guinea is the most ethno-linguistically diverse place on the planet. Its 3.5 million people speak 869 distinct languages.

2 Pirate Panache

Legendary and ruthless sea-raider "Black Bart" may win the award for the most prolific pirate, with more than 400 ships reportedly falling to his sword in the early 18th century. But Bart was much more civilized than history would have you believe. The Welsh-born Bartholomew Roberts (sound less tough now, doesn't he?) always wore a damask waistcoat, snappy breeches, and a dashing red feather in his cap. The refined Bart also drank only tea and water, commanded lights-out by 8 p.m., and had musicians play hymns for him on Sundays.

3 X Marks the 401(k)

When pirate icon Edward "Blackbeard" Teach met his Waterloo at Ocracoke Island (his pillaging hub off the coast of North Carolina) in 1718, his enemies confiscated 25 hogshead of sugar, 145 bags of cocoa, a barrel of indigo, and a bale of cotton. Not exactly the sacks full of rubies and sapphires the British Royal Navy was hoping for. When asked where the real treasure was, it's said he replied, "Only I and the devil know."

Since that time, beachcombers have donned Hawaiian-print shirts and scoured the Carolina coast with metal detectors—most likely in vain. Blackbeard's treasure is almost certainly more legend than fact. Pirates usually acquired their pieces of eight (Spanish silver coins), gold doubloons, and pricey jewels from black market trade of the coffee, tea, slaves, textiles, and medicines they stole from ships. But for all the talk of buried treasure, pirates weren't known for their retirement planning. They usually blew the money on women, booze, and gambling.

4 Playing the Parrot Card

Our modern-day image of a pirate usually comes fully outfitted with peg leg, eye patch, and parrot. Why? The stereotype comes directly from the fictional character of Long John Silver in Robert Louis Stevenson's *Treasure Island*. Silver's feathered sidekick, Captain Flint, was a nice touch, but it's doubtful pirates had pets. With long voyages and scanty rations, a parrot would have a made a better snack than companion.

5 Stealing Second

The Pittsburgh Pirates haven't always been named after the thieves of the high seas. Originally, the Major League club was known as the nature-loving Pittsburgh Alleghenies (after the mountain range in the eastern region of Pennsylvania). But in 1880, after stealing away second baseman Louis Bierbauer from the Philadelphia Athletics, a local newspaper called the team "a bunch of pirates." This suited them just fine, and they've been flying the Jolly Roger ever since.

In 1979 Japan offered new British PM Margaret Thatcher 20 "karate ladies" for protection at an economic summit. She declined.

SIX CANDIDATES FOR THE WORLD'S FIRST POP SONG

1 "Sumer Is Icumen In" (c.1239)

Why it might be the one: It didn't tell a story, or sing praise to God. Like most pop songs, it was about . . . nothing, really. Welcome to the *Seinfeld* of medieval music.

In medieval times, courts employed minstrels (or "jongleurs") to sing sagas or legends, as much to pass on information as for entertainment. These guys would bring their songs on the road, spreading them through the villages. But musical notation (in the West, at least) wasn't invented until around 1020, to make sure that every church parish was chanting the same tune. In the early days, most notated songs were hymns.

Heavy metal masters Black Sabbath originally went by a less imposing name: the Polka Tulk Blues Band.

Possibly the first major piece of non-hymnal music to find a mass audience was "Sumer is Icumen In," which predates the printing press by at least 150 years. After Johannes Gutenberg's invention came to England, however, it was published in all its glory. Here was a song in six parts (unheard of at the time), sung in an endless "round." Rather than praising God, it simply extolled the joys of summer, like so many later pop songs. "Sumer is icumen in," it began. "Lhude sing cuccu." (Or "Summer has arrived, loud sing the cuckoo.") Was it popular enough to be the first "pop" song? Maybe . . . but if we said "yes," this would be a really short list.

2 "Greensleeves" (c.1580)

Why it might be the one: One of the first songs to be printed as sheet music.

A few centuries before it was cheapened by ice cream vans and endless reruns of the *Lassie* TV series, this was possibly the first widely heard song in the English language, a love ballad with a melody as catchy as anything by the Beatles or Sara Bareilles. Strangely, it probably began life as a vigorous dance tune. It is often credited to Henry VIII, but while he was supposedly an accomplished musician, he probably can't claim this one. The words were first published around 1580 (some years after they were written).

3 "Home, Sweet Home" (1823)

Why it might be the one: Another new invention called the gramophone.
The simple lyrics written by John Howard Payne and hummable melody by Henry Bishop made this opera song a hit with the masses. But what really might give it the "first pop song" title is that, some 80 years later, it was one of the first songs to win major success on the gramophone, famously performed by at least three of the earliest recording stars: Australian diva Dame Nellie Melba, Italian "Queen of Song" Adelina Patti, and the "Swedish Nightingale," Jenny Lind.

When gramophone records were invented, short songs were slow to catch on, which is surprising, because they were ideal: early discs could hold only a few minutes of music. Yet even as late as 1910, more than three-quarters of records sold were classical pieces. Still, recorded music allowed a greater audience for music than ever before, no longer limited to households with a piano or a sight-reading singer.

"Oh! Susanna!" (1848)

Why it might be the one: A big hit (but we're not sure exactly how big).

If you thought that pop music was an American invention . . . you may be right. Pennsylvania-born Stephen Collins Foster's songs were inspired by (and often mistaken for) Negro spirituals, with their smoother and more accessible melodies than the intricate, opera-inspired tunes of the time. Though he published his first song, "Open Thy Lattice, Love," at age 18, "Oh! Susanna!" was his first major hit. Exactly how successful is difficult to say,

Tenor vocalist Luciano Pavarotti's standard contract required that there be no noise or "distinct smells" within the vicinity of the artist.

because song piracy was an issue even in the mid-19th century. More than 20 editions of the sheet music, mostly illegal, had spread all over the United States within three years. But despite the piracy, the publisher still made $10,000. (As a mere writer, Foster himself was given $100 for his troubles.)

"After the Ball" (1892)

Why it might be the one: The first million-seller—and this was before records!

The success of "After the Ball" was truly amazing. Before it was published, million-selling songs were unheard of. "After the Ball" sold five million copies within a year—as sheet music. The secret: a new(ish) concept called PR. Charles K. Harris, one of America's first songwriter-publishers, cannily promoted his song. In the United States, baritone J. Aldrich Libbey performed it at beer halls and theaters, in return for a share in the royalties. In Britain, it was a music-hall favorite. The mournful ballad also established Tin Pan Alley (a group of music publishers clustered around New York's Broadway) as the Mecca of popular song. Despite the detailed story told by the lyrics, the tune itself was simple enough. Harris couldn't even read music. "After the Ball" is his only song that anyone remembers, but that was enough for him to retire.

6 "My Gal Is a High Born Lady" (1896)

Why it might be the one: It signaled the birth of modern pop music . . . eventually.

The touring minstrel shows of the 19th-century, in which white singers would perform popular songs in blackface, are now dismissed as racist. But in a way, they were a compliment to black music. Despite their low social status, African-Americans were considered good musicians, partly due to their "sense of rhythm." Foster's black-inspired songs were, fittingly, made popular by minstrel groups. Even "After the Ball," inspired more by English ballads, was written for a minstrel show.

With Barney Fagan's now-forgotten "My Gal Is a High Born Lady," black music finally filtered into the mainstream, introducing a new, "boppier" style: ragtime. At the time, nobody knew how important this would be. But ragtime was the forerunner of jazz, rock 'n' roll, and almost every other major style of popular music in the next century. To an extent, the ragtime composers invented pop music as we know it. A Jewish composer, Irving Berlin, made his songwriting debut in 1911 by selling four songs in this style, all with "rag" or "ragtime" in the title (including the megahit "Alexander's Ragtime Band"). Not for the first time, a white man was spreading "black" music to the masses.

Before turning his attention to sports cars, Ferruccio Lamborghini built tractors.

NINE QUIRKY COLLEGE DONATIONS AND THE STRINGS ATTACHED

1 You Don't Bring Me Flowers

For years, Indiana University offered a scholarship with a strange condition: the recipient had to drive from Bloomington to Indianapolis once a year to put flowers on the donor's grave. The school gradually decided it was a bit much to ask a student to take a road trip to a stranger's headstone, though, so for 20 years it didn't enforce the requirement. Eventually the donor's attorney found out that the flowers weren't being placed, but rather than yank the gift he worked with the school to remove the clause from the bequest.

2 Auburn Goes to the Dogs (and Vice Versa)

When Miss Eleanor Elizabeth Ritchey, granddaughter of the founder of the Quaker State Oil Refining Company, died in 1968, she left Auburn University a generous gift of $2.5 million. She also gave the school something a bit more unusual: the responsibility for 150 dogs. Ritchey loved to adopt homeless dogs at her Florida ranch, so she made the large cash donation contingent on the school finding good homes for all 150 of her pooches. The cash was then earmarked for veterinary research.

3 Mystery Donor Opens a Giant Wallet

In 2009, colleges experienced an unprecedented rash of anonymous generosity. Colleges of all sizes around the country received letters from lawyers informing them of seven-figure anonymous donations. The only catch was that the donor wished to stay anonymous, and in some cases the giver required that the colleges sign a contract agreeing not to investigate the benefactor's identity. The donations, which ranged from $1 million all the way up to $10 million all went to schools that had female heads. Beyond that, though, the donor's identity and motives remained a mystery, even though he or she donated more than $70 million.

4 Bryn Mawr Goes on the Clock

Did Bryn Mawr need any new clocks in 1957? It didn't matter. They were getting one. Philadelphia physician Florence Chapman Child left the school $50,000 in her will if they would also agree to take her 150-year-old grandfather clock. The doctor stipulated that the school's administrators had to "install it in an appropriate place, keep it in proper condition and repair, make no changes in the fundamental appearance, and are not to have it electrified."

In 1928, a Canadian man left $568K to the Toronto mom who had the most kids in the next decade. Four split the prize—each had nine babies.

Bequest Puts Jocks on the Ropes

In 1907, fledgling Swarthmore College received a bequest that was estimated to be worth somewhere between $1 and $3 million. If the school wanted the cash, though, it would have to stop participating in intercollegiate sports. Swarthmore badly needed the cash—its entire endowment was only in the $1 million range—but in the end, the school turned down the gift and the sports survived.

Small Potatoes Lead to Big Cash

In 1950, the government had a surplus of potatoes and started looking for ways to get rid of the excess tubers. The Department of Agriculture decided to give the potatoes to Hiwassee College, a small Methodist school in eastern Tennessee. College president D. R. Youell told the government that he didn't want its charity, though. A short time later, the school received a $10,000 donation with a note praising the institution for taking a stand against "the dangerous trends toward socialism in our Government."

Students at Brigham Young University need a doctor's note to grow a beard.

7 College Profits From a Racist Will

When Dr. Jesse C. Coggins died in 1962, he left his estate to the Keswick nursing home so it could construct a new building. Nice gift, right? It would have been, but Coggins made a last-minute change to the will that stipulated that the building would house only white patients. In 1999, a court ruled that the racist stipulation effectively voided the gift and gave the entire estate—which had grown to $28.8 million—to the will's backup beneficiary, the University of Maryland Medical Center.

8 Donor Affects Fashion from Beyond the Grave

Radcliffe once received a piece of jewelry as a bequest. A nice gift, to be sure, but the late donor was a bit bossy. She wasn't just donating the piece of jewelry; she stipulated in the gift that the president of Radcliffe must wear the accessory.

9 . . . and for the College That Has Everything

The social philosopher left University College London a rather odd bequest in his will: his preserved, clothed body. No one's quite sure what Jeremy Bentham was getting at with this "gift," but since his 1832 death his clothed skeleton—topped with a wax model of Bentham's head—has been preserved in a wood-and-glass cabinet known as the Auto-Icon. It now resides at the school and is occasionally moved so Bentham can "attend" meetings. (Minutes of these meetings record his attendance as, "Jeremiah Bentham, present but not voting.")

Bentham didn't want the Auto-Icon to feature a wax head; for years he actually carried around the glass eyes he wanted used in his preserved face. However, the preservation process distorted his face, so the wax replica had to stand in. For many years Bentham's real head sat between his feet in the Auto-Icon, but it was such a target for pranksters that it eventually had to be locked away.

Texas A&M University has its own official greeting: "Howdy."

THE ORIGINS OF EIGHT SUPERSTITIONS (SEVEN WOULD HAVE BEEN LUCKIER)

Knocking on Wood

After bragging, it's a common gesture to tap on something wooden—a chair, a door, or (for the remaining few who still insist it's funny) your own head—in order to show humility and the hope that circumstances remain the same. But a few thousand years ago, our ancestors were thumping on trees. To them, trees represented the homes of the gods. So whenever a favor was needed, a tree was touched, and, when the favor was received, the tree was touched again in appreciation. Centuries later, medieval Christians wore pendant wooden crosses, like the one Christ was crucified on, which they tapped in reverent apology whenever they caught themselves boasting to remind themselves of their humility.

Ancient Egyptians swore on garlic the same way people today take oaths on the Bible.

2 Tossing Coins into Fountains

Nope, this tradition did not start at your local mall. It actually began in ancient times, when it was believed that spirits lived in the bottom of wells. The thinking was that if one made a wish and then tossed money into the well, the spirits would be pleased and would grant the wish. However, there is another story associated with the origin of this practice.

According to Greek mythology, mortals would bury their dead with an obol (a very small coin) under the deceased's tongue. At the time, Greeks believed that the souls of the dead needed money in order to cross the river Styx, which served as the crossroads between the living and the dead. If they had money, Charon, the greedy ferryman, would send them across to a peaceful afterlife; if they didn't, they wouldn't be allowed to cross.

3 Throwing Rice at a Wedding

Throwing rice onto the bride and groom after they've said their vows stems from ancient Hindu and Chinese religions. As a primary food source, rice became a symbol of health, prosperity, and fertility. It was also thrown in hopes of appeasing any malicious spirits that may have been lurking nearby.

4 Walking under a Ladder

This superstition most likely originated in ancient Asiatic countries, where a criminal who was sentenced to death (as most criminals were) was hanged from a ladder propped against a tree. And since it was believed that death was contagious, people were forbidden to walk underneath ladders for fear that they'd "catch death." Later, the onset of Christianity helped to spread this conviction, as early paintings of Jesus' crucifixion depicted a ladder leaning against the cross with Satan standing underneath. Being caught under a ladder therefore became associated with being in Satan's territory.

5 Throwing Salt over Your Shoulder

Salt was pretty valuable to the ancient Greeks. Not only did they use it to preserve food, but many Greeks were actually paid in salt (called their salarium, from which the word salary stems). Because of its importance, they concluded that salt could also protect them from evil spirits. But why throw it over your shoulder? Well, the Greeks believed that spilling salt was a sign from your guardian spirit that dark apparitions were on the prowl. As to which shoulder these demons were lurking over, it was commonly thought that positive spirits congregated on the right side of the body, while negative ones resided on the left. Many thought that the pinch of salt they were throwing over their left shoulder served to bribe the demons, while others were just hoping some would land in their eyes and blind them.

The U.S. Mint reported a loss of $42.6 million on the production of pennies and nickels in 2010.

6 Fear of Breaking a Mirror

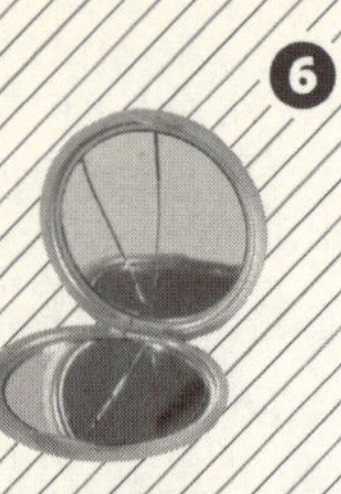

Long before breakable mirrors existed, a shiny surface was considered a tool of the divine. In addition, early man (believing that his reflection in still water was an image of his soul) feared having his reflection harmed in any way. About the first century BCE, the Romans, who believed that life renewed itself every seven years, reasoned that breaking a mirror would summon seven years of bad luck.

7 Fear of Black Cats

A black cat crossing your path is supposed to bring bad luck, a throwback to the Middle Ages when it was believed that this creature was the companion and mascot of a coven and that it changed into a witch or Satan after seven years of service. Therefore, it stood to reason that a black cat crossing one's path might be a witch or the devil in disguise.

8 Paraskevidekatriaphobia

Yup, a fear of Friday the 13th is now classified as its very own phobia. Historically, both Friday and the number 13 have had unlucky associations, so it only stood to reason that a fear of the combination of the two would arise. The most commonly cited explanation for fearing Friday the 13th is that there were 13 people at the Last Supper, one of whom betrayed Jesus, and that the Crucifixion fell on a Friday. During the days of the Druids, Friday was the night of the witches' Sabbath, where they always gathered in covens of 12 + 1—the thirteenth member believed to be the devil.

Editor's note: Because most superstitions have existed for so many centuries, multiple stories of their origins exist. The origins listed in this article are simply the most prominent and well-researched derivations.

The only number whose letters are in alphabetical order is 40 (f-o-r-t-y).

TICKET TO RIDE

FIVE THINGS YOU DIDN'T KNOW ABOUT THE LOTTERY

1 Lotteries of Yore (They're Older Than You Think!)

Lotteries have been around as long as arithmetic. According to the Bible, God ordered Moses to use a lottery to divvy up land along the River Jordan. That story came from the Book of Numbers (naturally), but lotteries are also mentioned in Joshua, Leviticus, and Proverbs. The concept can also be traced back to China, where a warlord named Cheung Leung came up with a numbers game (today known as keno) to persuade citizens to help pay for his army.

Other famous lotteries? Augustus Caesar authorized one to raise money for public works projects in Rome. The Chinese used one to help finance the Great Wall. And in 1466, in what is now the Belgian town of Bruges, a lottery was created to help the poor—a fund-raising effort they've been doing ever since.

2 The Founding Fathers Took Their Chances

Displaying every astute politicians' aversion to direct taxation, early American leaders often turned to lotteries to raise a buck or two. John Hancock organized several of them, including one to rebuild Boston's fire-stricken Faneuil Hall. Ben Franklin used them during the Revolutionary War to purchase a cannon for the city of Philadelphia. George Washington ran a lottery to pay for the construction of a road into the wilds of western Virginia. And Thomas Jefferson wrote of lotteries, "far from being immoral, they are indispensable to the existence of Man." Of course, when Jefferson wrote that, he was trying to convince the Virginia legislature to let him hold a lottery to pay off his debts.

The odds of hitting the Powerball jackpot are roughly 1 in 195 million. For perspective, the odds of being hit by lightning *twice* are just 1 in 39 million.

Louisiana: A Whole Lotto Love

By the end of the Civil War, lotteries in America had such bad reputations, they'd been banned in most states. But not in Louisiana, where a well-bribed legislature in 1869 gave an exclusive charter to a private firm called the Louisiana Lottery Company, which sold tickets throughout the country. For 25 years, it raked in millions of dollars while paying out relatively small prizes and contributing chump change to a few New Orleans charities.

Finally, in 1890, Congress passed a law banning the sale of lottery tickets through the mail, and eventually, all multistate lottery sales were banned. What's a corrupt U.S. company to do? Move offshore, of course! The Louisiana Lottery moved its operations to Honduras, and America was lottery-free until 1963, when New Hampshire started the lottery cycle anew.

4 Inaction Jackson: Lottery's Biggest Loser

Clarence Jackson's luck began to run out on Friday the 13th in October 1995, when the Connecticut Lottery picked the numbers on Jackson's lotto ticket. He and his family won $5.8 million, only they didn't know it. In fact, they didn't find out until 15 minutes before the one-year deadline to claim the prize, despite a whole slew of lottery ads seeking the winner.

Jackson, a 23-year-old who'd taken over the family's struggling office-cleaning business from his ailing father, didn't make it in time, and lottery officials rejected the claim. In 1997, the Connecticut House of Representatives voted to award Jackson the prize, but the state senate refused to go along. Up until 2004, Jackson was still making annual attempts to convince the legislature otherwise, and still losing.

5 And Some Other Jackson: Its Biggest "Winner"

Andrew Jackson "Jack" Whittaker was already wealthy when he won the multistate Powerball lottery in December 2002. A millionaire contractor from West Virginia, Whittaker became the biggest single lottery winner in history after snagging a $314.9 million jackpot. But the dough seemed to carry more curses than the Hope Diamond. When Jack decided to collect his winnings in a $170.5 million lump sum, rather than payments over 20 years, it wasn't the only lump coming his way. Whittaker was robbed three times, once of more than $500,000 at a strip club. He was also sued for assault, arrested multiple times for drunk driving, and even booked for getting into a bar fight.

The sum of all the numbers on a roulette wheel is 666.

WORDS TO SING BY

FIVE SONGS THAT SOUND LIKE THEY'RE ABOUT WOMEN (BUT AREN'T)

It may undercut the romance a bit, but in some cases knowing that the song isn't actually about a girl makes it a lot cooler.

1 "Cracklin' Rosie" by Neil Diamond

She's not a girl—she's a bottle of wine. Neil Diamond told *Rolling Stone* that he got the idea from a tribe in Canada with a much higher male population than female population. When the guys with girlfriends all went out on dates, the bachelor guys got together and drank homemade hooch. Cracklin' Rosie is actually pretty decent wordplay when you realize the song's boozy origins: "crackling" is used in the wine world to describe a wine that's lightly sparkling. You can actually buy a crackling rosé.

The title of the Paul Simon song "Mother and Child Reunion" came from a chicken and egg dish on a Chinese food menu in New York City.

"Elenore" by the Turtles

If you've ever listened to that song and thought the line "you're my pride and joy etcetera" felt like a pretty poor attempt at lyrics, well, you're right. "Elenore" wasn't a girl, but an anti-love letter to the Turtles' record label. The label was demanding a hit just like the previous year's "Happy Together," so the band slapped together a song they felt was insanely clichéd, happy-go-lucky, and crappy. To everyone's surprise, it was a big hit.

3 "Mony, Mony" by Tommy James and the Shondells.

The band had the song, but it needed a two-syllable girl's name to stick in the lyrics. That's when Tommy James happened upon a Mutual of New York Insurance Company sign that had a dollar sign in the middle of the "O" and noticed that its acronym was MONY. He and his songwriting partner had a laugh about what a great name MONY was for such a company and ended up calling their fictional girl "Mony."

4 "Sexy Sadie" by the Beatles

Sexy Sadie was really Maharishi Mahesh Yogi. The song was about how they were disillusioned with the holy man after he hit on Mia Farrow and some of the other girls studying with him. Mia Farrow and most of the group later felt bad for making these accusations, feeling they had probably misinterpreted his actions.

"Martha, My Dear" by the Beatles

Paul McCartney's girlfriend Jane Asher may have inspired the lyrics to this one—he says they "probably" were based on her—but the name comes from his Old English Sheepdog.

The tallest jockey on record is 7'7" ex-NBA center Manute Bol. The Indiana Horse Racing Commission licensed Bol to race in a 2003 charity event.

SEALED WITH A KISS

THE FIVE MOST IMPORTANT SMOOCHES IN THE UNIVERSE

1 The Most Iconic Kiss in History

On August 14, 1945, thousands of men and women embraced one another in New York City's Times Square to celebrate victory over Japan. But two people—a sailor and a nurse—locked lips at just the right moment and became larger than life. More than a dozen men and at least three women claim to be the kissers in Alfred Eisenstaedt's photograph.

Of the men, our favorite is George Mendonça, a Rhode Island fisherman and World War II navy recruit, who claims he grabbed the strange nurse and kissed her right in front of his girlfriend. In fact, Mendonça says his girlfriend, now his wife, is in the background of the photo.

While the mystery will probably never be solved, Alfred Eisenstaedt has left us with a juicy backstory. In his autobiography, the famed photographer writes that he followed around a sailor who moved through the crowd, kissing anything wearing a skirt. When the sailor hit on a nurse whose white dress contrasted nicely with his dark suit, Eisenstaedt snapped the shot. But he failed to get their names.

Coincidentally, another photographer, Victor Jorgensen, took the same shot from a slightly different angle and also forgot to get the subjects' names. Jorgensen's version ran in the next day's *New York Times*, but as a working military photographer at the time, he didn't own the rights to his work. So while Eisenstaedt received glory and royalty checks for his image, Jorgensen simply got a nice clipping to hang on his fridge.

2 The Kiss That Proved No Means No

Gentlemen, a word: When a lady rejects your advances, you'd do best to listen. Take, for example, the story of Thomas Saverland, an English gentleman who was at a party in 1837 and, as a joke, kissed Miss Caroline Newton by force. In response, she bit off a chunk of his nose.

Saverland took her to court, where the judge found his case more hilarious than harrowing. The judge ruled, "When a man kisses a woman against her will, she is fully entitled to bite off his nose, if she so pleases." A smart-mouthed barrister then added, "and eat it up, if she has a fancy that way."

The average person spends about 20,160 minutes of their life kissing. That's enough to burn 524,160 calories, or about 970 Big Macs.

The Kiss That Said "Welcome to America!"

At the turn of the 20th century, immigration processing at Ellis Island was quite an ordeal. Ladies traveling alone and anyone with less than $20 in their pockets had to wait for a sponsor or family member to meet them. If no one was there to greet you, you were sent back.

The process was further complicated by the fact that immigrants couldn't go down to the pay phone and call Aunt Bertha when they landed. Instead, when relatives heard that the right ship had docked, they trucked over to Ellis Island and waited desperately by the Kissing Post—a giant wooden column just outside the room where the final stages of immigration took place. Ellis Island staffers gave the Kissing Post its name because families and lovers were generally swept up in emotion as they reconnected with their long losts.

4 The Eskimo Kiss: A Tale Taller than the Abominable Snowman

Popular wisdom claims that Eskimos rub noses because kissing on the lips would cause their mouths to freeze together. Not only is this completely untrue, but Eskimos don't rub noses at all.

The myth of the Eskimo kiss was created by Hollywood in an early "documentary" called *Nanook of the North*, which took America by storm in 1922. To film it, director Robert J. Flaherty recorded real Inuits in the Arctic. However, in order to accommodate the huge, awkward cameras of the day, he staged all the scenes and built a three-sided igloo for interior shots. Nanook, the main character, wasn't really named Nanook, and the women playing his wives weren't really his wives.

As for the term "Eskimo kiss," that too was constructed by Flaherty to explain how one of the wives was nuzzling her baby. In actuality, the woman was giving her baby a kunik, an expression of affection in Inuit culture. Typically in kuniks, adults press the sides of their noses against the cheeks of their babies and breathe in their scent. Who kuniks whom differs from culture to culture, but it's never a romantic gesture. Inuits kiss on the lips, just like everyone else.

5 The Kiss That Could Send You to Jail

In the city of Guanajuato, Mexico, there's a smooching spot called El Callejon del Beso, or the Alley of the Kiss. According to local legend, the alley was once the final scene of a tragic love affair. A young woman and her lover were meeting there to run away together, but when her father discovered them, he stabbed his daughter in the heart. As she lay dying, her lover kissed her hand for the last time, and the alley got its name. Today, it's said that anyone who kisses there will have seven years of happiness.

Thanks to its romantic history, the alley has become a popular tourist attraction, although that's starting to change. On January 20, 2009, the ultraconservative mayor of Guanajuato authorized a new municipal ordinance cracking down on public displays of affection. If he has his way, lip-locking in the open will carry with it a fine of $100 and up to 36 hours in jail.

Tweety Bird was originally pink. Animators added yellow feathers in response to criticism that the bird was naked.

NERD PILGRIMAGES

FIVE ODDLY SPECIFIC MUSEUMS PRESERVING OUR HISTORY

The Spam Museum

If the on-site "wall of SPAM" is any indication, a tour through the SPAM Museum in Austin, Minn., is guaranteed fun for the whole canned-pork-loving family. SPAM's parent company, Hormel Foods, opened the establishment in 2001 to the tune of almost 5,000 cans of SPAM. One of the main attractions is a scale model of a SPAM plant, where visitors can don white coats and hairnets while pretending to produce America's favorite tinned meat.

Queen Elizabeth II officially owns all of the mute swans that live in the Thames River.

National Museum of Funeral History

It's pretty hard to argue with the motto "Any Day Above Ground is a Good One." So goes the backhanded optimism of the National Museum of Funeral History, a Houston facility that opened in 1992. Visitors are treated to exhibits that include a Civil War embalming display and a replica of a turn-of-the-century casket factory. In addition, the museum boasts an exhibit of "fantasy coffins" designed by Ghanaian artist Kane Quaye. These moribund masterpieces include a casket shaped like a chicken, a Mercedes-Benz, a shallot, and an outboard motor. According to Quaye, his creations are based on the dreams and last wishes of his clients, which—let's be honest—really makes you wonder about the guy buried in the shallot.

The Hobo Museum

If you're bumming around but looking for a good time, be sure to take a load off in Britt, Iowa, at the Hobo Museum, which details the history and culture of tramps. Bear in mind, though, that the museum kind of, well, slacks on hours and is only open to the public during the annual Hobo Convention. Luckily, tours can be arranged by appointment any time of year. Of course, if you're interested in the Hobo Convention, lodging is available all over the area, but it's a safe bet that most of your compatriots will be resting their floppy hats at the "hobo jungle," located by the railroad tracks. Both the event and the museum are operated by the Hobo Foundation, which—incidentally—also oversees the nearby Hobo Cemetery, where those who have "caught the westbound" are laid to rest.

4 The Museum of Questionable Medical Devices

Take two trips to the Museum of Questionable Medical Devices and call us when you've lost all faith in the medical profession. Thanks to curator Bob McCoy (who has donated the collection to the Science Museum of Minnesota), those in search of history's quack science can find what they're looking for in the St. Paul tourist attraction, whether it's a collection of 19th-century phrenology machines or some 1970s breast enlargers. If you make the trip, be sure to check out the 1930s McGregor Rejuvenator. This clever device required patrons to enclose their bodies, sans head, in a large tube where they were pounded with magnetic and radio waves in attempts to reverse the aging process.

5 The Trash Museum

Mom wasn't kidding when she said one man's trash is another man's treasure. At the Trash Museum in Hartford, the Connecticut Resources Recovery Authority (CRRA) turns garbage into 6,500 square feet of pure recycling entertainment! Tour the Temple of Trash or visit the old-fashioned town dump. And for your recycler-in-training, head down the street to the Children's Garbage Museum, where you can take an educational stroll through the giant compost pile, get a glimpse of the one-ton Trash-o-saurus, or enjoy the company of resident compost worms.

The taxicab gets its name from the taximeter, a device that measures the distance and time a car has traveled.

PHOTO CREDITS

p. II ajt/Shutterstock.com
p. III Fotoline/Shutterstock.com
p. IV Feng Yu/Shutterstock.com
p. V Ivonne Wierink/Shutterstock.com
p. VI Barbara J. Petrick/Shutterstock.com
p. VII Zany Zeus/Shutterstock.com
p. VIII Joe Gough/Shutterstock.com
p. IX Anthony DiChello/Shutterstock.com
p. X Serg64/Shutterstock.com
p. 1 Warren Price Photography; Alexander Ishchenko; R. Carner; moritorus/Shutterstock.com
p. 2 Simon Krzic/Shutterstock.com
p. 3 Pincasso/Shutterstock.com
p. 5 Jesse Kunerth/Shutterstock.com
p. 6 Nikolay Petkov/Shutterstock.com
p. 8 Arunas Gabalis/Shutterstock.com
p. 10 Gary Yim/Shutterstock.com
p. 11 Carsten Reisinger/Shutterstock.com
p. 12 Givaga/Shutterstock.com
p. 13 Teresa Azevedo/Shutterstock.com
p. 15 Kristof Degreef/Shutterstock.com
p. 17 PeJo/Shutterstock.com
p. 18 Rostislav Glinsky/Shutterstock.com
p. 19 Stevan Kordic/Shutterstock.com
p. 20 diak/Shutterstock.com
p. 21 Gino Santa Maria/Shutterstock.com
p. 22 Kletr/Shutterstock.com
p. 24 Travis Manley/Shutterstock.com
p. 25 Gino Santa Maria/Shutterstock.com
p. 26 John Kroetch/Shutterstock.com
p. 27 Ultrashock/Shutterstock.com
p. 28 Marty Ellis/Shutterstock.com
p. 29 MIT
p. 30 F.G.C./Shutterstock.com
p. 32 JCVStock/Shutterstock.com
p. 33 markhiggins/Shutterstock.com
p. 34 Martin Wall/Shutterstock.com
p. 35 Mark Aplet/Shutterstock.com
p. 36 Jiri Vaclavek/Shutterstock.com
p. 37 Jens Stolt/Shutterstock.com
p. 38 Ariene/Shutterstock.com
p. 39 Dudarev Mikhail/Shutterstock.com
p. 40 Fotoline/Shutterstock.com
p. 42 c. /Shutterstock.com
p. 43 Bettmann/Corbis
p. 44 Cliparea | Custom media/ Shutterstock.com
p. 46 Losevsky Pavel/Shutterstock.com
p. 47 Victorian Traditions/ Shutterstock.com
p. 48 topal/Shutterstock.com
p. 49 Karen Gentry/Shutterstock.com
p. 50 Louie Psihoyos/Science Faction/ Corbis
p. 51 Matt Valentine/Shutterstock.com
p. 52 Frank Conlon/Star Ledger/Corbis
p. 53 AnRo brook/Shutterstock.com
p. 54 Rod Ferris/Shutterstock.com
p. 55 Rob Byron/Shutterstock.com
p. 57 Colour/Shutterstock.com
p. 58 Sergio Dorantes/Sygma/Corbis
p. 59 Brandon Jennings/Shutterstock.com
p. 60 Tatiana Popova/Shutterstock.com
p. 62 Yco/Shutterstock.com
p. 63 ajt/Shutterstock.com
p. 64 eyespeak/Shutterstock.com
p. 65 Bettmann/Corbis
p. 66 ronstik/Shutterstock.com
p. 68 rorem/Shutterstock.com
p. 69 jannoon028/Shutterstock.com
p. 70 Dmitry Fisher/Shutterstock.com
p. 71 Photo NAN/Shutterstock.com
p. 72 Carolina K. Smith, M.D./ Shutterstock.com
p. 73 fotohunter/Shutterstock.com
p. 74 Iwona Grodzka/Shutterstock.com
p. 76 africa 924/Shutterstock.com
p. 77 Studio DMM Photography, Designs & Art/Shutterstock.com
p. 78 Luis Santos/Shutterstock.com
p. 80 @erics/Shutterstock.com
p. 81 Shell114/Shutterstock.com
p. 82 James Steidl/Shutterstock.com
p. 83 Bruce Works/Shutterstock.com
p. 84 Filip Fuxa/Shutterstock.com
p. 86 Laurin Rinder/Shutterstock.com
p. 87 Eviakhov Valeriy/Shutterstock.com
p. 88 Bettmann/Corbis
p. 89 Rafa Irusta/Shutterstock.com
p. 90 abadesign/Shutterstock.com
p. 91 Steve Collender/Shutterstock.com
p. 92 Dirk Ercker/Shutterstock.com
p. 93 Péter Gudella/Shutterstock.com
p. 94 digitalreflections/Shutterstock.com
p. 96 Bettmann/Corbis
p. 97 Yannis Ntousiopoulo/ Shutterstock.com
p. 98 Giampiero Sposito/Reuters/Corbis
p. 99 Jiri Hera/Shutterstock.com
p. 100 xstockerx/Shutterstock.com
p. 101 Dimitri Iundt/TempSport/Corbis
p. 102 Philip Lange/Shutterstock.com
p. 104 Bettmann/Corbis
p. 105 deepspacedave/Shutterstock.com
p. 107 3drenderings/Shutterstock.com
p. 108 Chas/Shutterstock.com
p. 109 Bettmann/Corbis
p. 111 Bjorn Heller/Shutterstock.com
p. 112 Larisa Lofitskaya/Shutterstock.com
p. 113 MarFot/Shutterstock.com
p. 114 Maksymilian Skdik/Shutterstock.com
p. 115 Dave Nevodka/Shutterstock.com
p. 116 Stringer/Italy/Reuters/Corbis
p. 117 L. L. Masseth/Shutterstock.com
p. 119 max777/Shutterstock.com
p. 120 Faiz Zaki/Shutterstock.com
p. 121 Bialy, Doratai Bogdan/the food passionates/Corbis
p. 122 jirasaki/Shutterstock.com
p. 124 EuTouch/Shutterstock.com
p. 125 Bob Krist/Corbis
p. 126 Alaettin Yildirim/Shutterstock.com
p. 129 Matt Valentine/Shutterstock.com
p. 130 nito/Shutterstock.com
p. 131 Paul Prescott/Shutterstock.com
p. 132 Constantinos/Shutterstock.com
p. 135 Marc Dietrich/Shutterstock.com
p. 136 Bettmann/Corbis
p. 137 zoommer/Shutterstock.com
p. 139 Dja65/Shutterstock.com
p. 140 Katherine Welles/Shutterstock.com
p. 141 Cre8tive Images/Shutterstock.com
p. 142 Val Lawless/Shutterstock.com
p. 143 Greg Kushmerek/Shutterstock.com
p. 144 Troy Lochner/Shutterstock.com
p. 146 Bobby Deal/RealDealPhoto/ Shutterstock.com
p. 149 Arti_Zav/Shutterstock.com
p. 150 Suzanne Tucker/Shutterstock.com
p. 151 Morgan Lane Photography/ Shutterstock.com
p. 152 Stephen Coburn/Shutterstock.com
p. 153 Smailes Alex/Corbis Sygma
p. 154 Bara22/Shutterstock.com
p. 155 Andrea Skjold/Shutterstock.com
p. 156 Laura Stone/Shutterstock.com
p. 157 Ken Tannenbaum/Shutterstock.com
p. 158 Myotis/Shutterstock.com
p. 160 Picsfive/Shutterstock.com
p. 161 titelio/Shutterstock.com
p. 162 Laurent Dambies/Shutterstock.com
p. 163 CinemaPhoto/Corbis
p. 165 Iakov Filimonov/Shutterstock.com
p. 167 Scott Rothstein/Shutterstock.com
p. 168 Bettmann/Corbis
p. 169 nutech21/Shutterstock.com
p. 171 Ioannis Pantzi/Shutterstock.com
p. 172 Petoo/Shutterstock.com
p. 173 Pahknyushcha/Shutterstock.com
p. 174 Sunset Boulevard/Corbis
p. 175 Vartanov Anatoly/ Shutterstock.com
p. 177 Anita Patterson Peppers/ Shutterstock.com
p. 178 James E. Knopf/Shutterstock.com
p. 179 Jose Alberto Tejo/Shutterstock.com
p. 180 Tischenko Irina/Shutterstock.com
p. 181 UltraOrto, S.A./Shutterstock.com
p. 182 J. Helgason/Shutterstock.com
p. 183 Fotoline/Shutterstock.com
p. 184 Tinydevil/Shutterstock.com
p. 185 Karen Hadley/Shutterstock.com
p. 186 Ingvar Bjork/Shutterstock.com
p. 187 ref348985/Shutterstock.com
p. 188 Elnur/Shutterstock.com
p. 189 Luba V Nel/Shutterstock.com
p. 190 aquatic creature/Shutterstock.com
p. 192 Thomas M Perkins/Shutterstock.com
p. 193 Doug Matthews/Shutterstock.com
p. 194 Olga Miltsova/Shutterstock.com
p. 196 JoLin/Shutterstock.com
p. 197 Mario Tarello/Shutterstock.com
p. 198 Corbis
p. 199 jathys/Shutterstock.com
p. 200 Steve Collender/Shutterstock.com
p. 201 joppo/Shutterstock.com
p. 202 Doug DeNeve/Shutterstock.com
p. 203 Kozoriz Yuriy/Shutterstock.com
p. 204 Khz/Shutterstock.com
p. 205 Keith Gentry/Shutterstock.com
p. 206 solaris_design/Shutterstock.com
p. 208 saragosa69/Shutterstock.com
p. 209 David Davis/Shutterstock.com
p. 210 Andy Dean Photography/ Shutterstock.com
p. 211 Bruce Works/Shutterstock.com
p. 213 kzww/Shutterstock.com
p. 214 tuanyick/Shutterstock.com
p. 215 Gelpi/Shutterstock.com
p. 216 Aksenova Natalya/Shutterstock.com
p. 217 rcphoto/Shutterstock.com
p. 218 Erik Kabik/Retna Ltd./Corbis
p. 219 Jiri Vaclavek/Shutterstock.com
p. 221 James Steidl/Shutterstock.com
p. 222 Rob Byron/Shutterstock.com
p. 223 AISPIX/Shutterstock.com
p. 224 Olemac/Shutterstock.com
p. 226 wonderisland/Shutterstock.com
p. 227 Gene Blevins/LA Daily News/ Corbis
p. 228 Erik Lam/Shutterstock.com
p. 229 Zeljko Radojko/Shutterstock.com
p. 230 Cosmin Manci/Shutterstock.com
p. 231 Suzanne Tucker/Shutterstock.com
p. 232 Tom Grundy/Shutterstock.com
p. 234 Sergey Khachatryan/ Shutterstock.com
p. 235 J. Helgason/Shutterstock.com
p. 237 magicoven/Shutterstock.com
p. 238 magnola/Shutterstock.com
p. 239 atoss/Shutterstock.com
p. 240 Jiri Hera/Shutterstock.com
p. 241 K Chelette/Shutterstock.com
p. 242 Photsani/Shutterstock.com
p. 243 Murat Baysan/Shutterstock.com
p. 244 Gerald Bernard/Shutterstock.com
p. 245 Paul Edmonson/Spaces Images/ Corbis
p. 246 Gordon Saunders/Shutterstock.com
p. 247 John Springer Collection/Corbis
p. 248 Bogdan Wankowicz/Shutterstock.com
p. 249 carl ballou/Shutterstock.com
p. 250 JaguarKo/Shutterstock.com
p. 252 Kinetic Imagery/Shutterstock.com
p. 253 farres/Shutterstock.com
p. 254 James Steidl/Shutterstock.com
p. 256 scherbet/Shutterstock.com
p. 258 Miles Luzanin/Shutterstock.com
p. 259 Klaus Schnitzer/Transtock/Corbis
p. 260 JR Trice/Shutterstock.com
p. 261 Bettmann/Corbis
p. 262 Danny Smythe/Shutterstock.com
p. 263 Hannes Eichinger/Shutterstock.com
p. 264 jokerpro/Shutterstock.com
p. 266 tatiana sayig/Shutterstock.com
p. 267 Rey Kamensky/Shutterstock.com
p. 268 Bettmann/Corbis
p. 269 David J. & Janice L. Frent Collection/ Corbis
p. 270 Aaron Kohr/Shutterstock.com
p. 271 Hulton-Deutsch Collection/Corbis
p. 272 Dimity Kolmakov/Shutterstock.com
p. 275 daseaford/Shutterstock.com
p. 276 Karen Hildebrand Lau/ Shutterstock.com
p. 277 James Steidl/Shutterstock.com
p. 278 Petoo/Shutterstock.com
p. 279 Jasper Lenselink Photography/ Shutterstock.com
p. 280 Jeff Banke/Shutterstock.com
p. 281 timy/Shutterstock.com
p. 282 Sergey Lavrentev/Shutterstock.com
p. 283 Nathan Holland/Shutterstock.com
p. 284 Ghoerghe Bunescu Bogdan Mircea/Shutterstock.com
p. 285 Tatiana Popova/Shutterstock.com
p. 287 Perry Correll/Shutterstock.com
p. 288 David Burden/Shutterstock.com
p. 289 Stephen Coburn/Shutterstock.com
p. 290 The Mariners' Museum/Corbis
p. 291 magnola/Shutterstock.com
p. 292 Petrov Artem/Shutterstock.com
p. 293 Onur Ersin//Shutterstock.com
p. 295 Elpis Ioannidis/Shutterstock.com
p. 297 Dedhyukhin Dmitry/ Shutterstock.com
p. 298 Ivan Cholakov Gostock.net/ Shutterstock.com
p. 299 oknoart/Shutterstock.com
p. 300 Karam Miri/Shutterstock.com
p. 301 Belova Larissa/Shutterstock.com
p. 302 Nils Z/Shutterstock.com
p. 303 dpaint/Shutterstock.com
p. 304 Rafa Irusta/Shutterstock.com
p. 306 Andrew Buckin/Shutterstock.com
p. 307 Joe Gough/Shutterstock.com
p. 308 Coprid/Shutterstock.com

English speakers say "cheese" when smiling for a photo, but in Chinese, it's "eggplant" and in Spanish, it's "potato."